Daniel Harrison Jacques

The Barn-Yard

A manual of cattle, horse and sheep husbandry; or, How to breed and rear the various species of domestic animals: embracing directions for the breeding, rearing, and general management of horses, mules, cattle, sheep, swine and poultry

Daniel Harrison Jacques

The Barn-Yard
A manual of cattle, horse and sheep husbandry; or, How to breed and rear the various species of domestic animals: embracing directions for the breeding, rearing, and general management of horses, mules, cattle, sheep, swine and poultry

ISBN/EAN: 9783337329297

Printed in Europe, USA, Canada, Australia, Japan

Cover: Foto ©berggeist007 / pixelio.de

More available books at **www.hansebooks.com**

THE BARN-YARD:

A MANUAL

OF

Cattle, Horse and Sheep Husbandry;

OR, HOW TO BREED AND REAR

THE VARIOUS SPECIES OF DOMESTIC ANIMALS:

EMBRACING

DIRECTIONS FOR THE BREEDING, REARING, AND GENERAL MANAGEMENT OF HORSES, MULES, CATTLE, SHEEP, SWINE AND POULTRY; THE GENERAL LAWS, PARENTAGE, AND HERIDITARY DESCENT, APPLIED TO ANIMALS, AND HOW BREEDS MAY BE IMPROVED; HOW TO INSURE THE HEALTH OF ANIMALS; AND HOW TO TREAT THEM FOR DISEASES WITHOUT THE USE OF DRUGS;

WITH A

Chapter on Bee-Keeping.

BY D. H. JACQUES,

AUTHOR OF "THE HOME," "THE GARDEN," "THE FARM," "HOW TO WRITE," "HOW TO DO BUSINESS," ETC.

Our power over the lower animals, if rightly exercised, redounds to their elevation and happiness no less than to our convenience and profit.—THE AUTHOR.

REVISED EDITION.

New York:
GEO. E. & F. W. WOODWARD,
No. 37 PARK ROW, OFFICE OF "THE HORTICULTURIST."
1866.

Entered, according to Act of Congress, in the year 1866, by

D. H. JACQUES,

In the Clerk's Office of the District Court of the United States for the Southern District of New York.

PREFACE.

We commenced this little manual with the intention of making the most useful compilation possible, within the space allowed us, from the great number of larger works on the subjects treated to which we had access. In the progress of our work, however, we found occasion to depart, in some degree, from our original plan, and introduce more new matter and re-write and condense more that is, in substance, derived from others, than we at first intended; but our claims on the score of originality will not be large. If the matter and arrangement of our book shall prove acceptable to the public, and serve the purposes intended, we shall be satisfied. The humble merit of having presented, in an attractive and available form a mass of useful information, practical hints, and valuable suggestions, on a number of important topics, is all that we purpose to insist upon. This the great public, for whose good we have labored, will, we are sure, readily accord to us.

We have given credit in the body of the work, whenever practicable, to the authors from whom we have derived aid in the various departments of our labor; but we here gladly make an additional record of our indebtedness to the works of Youatt, Martin, Stuart, Randall, Wingfield, Dixon, Bement, Browne, Quimby, etc. The *Country Gentleman*, the *American Agriculturist*, the *Southern Cultivator*, and other agricultural papers, have been examined with satisfaction to ourselves and with profit to our readers.

We have endeavored to make our little work thorough and reliable, so far as it goes, and to give the largest possible amount of useful information that can be condensed into so small a number of pages. We have occupied a large field, we are aware, and can not hope to have been so full on all points as many readers will desire. We have not aimed, of course, to render the larger works on the special topics to which our chapters are devoted unnecessary. We hope rather to create a demand for them; but there are thousands whom this little manual will furnish with all the information they desire on the subjects on which it treats, and on whom the details with which the larger and more expensive works are filled would be thrown away. To such, in an especial manner, we commend it, hoping that it will not wholly fail to meet their expectations.

CONTENTS.

I.—THE HORSE.

A Historical Sketch—Range of the Horse in Reference to Climate—Effects of Climate and Food—Varieties or Breeds—The Race-Horse—Origin and Characteristics—Half-bred Horses—The Arabian Horse—Wonderful Genealogies—Description—The Arabian "Tartar"—The Morgan—Opinions in Reference to the Morgans—Sherman Morgan—The Canadian Horse—The Norman—"Louis Philippe"—The Cleveland Bay—The Conestoga—The Clydesdale Horse—The Virginian—Wild Horses—American Trotting Horses—Points of a Horse Illustrated—Color, and what it Indicates—Common Terms Denoting the Parts of a Horse—Stables—Stables as they are—Situation of Stables—Size—Windows—Floors—Draining—Racks and Mangers—Ventilation of Stables—Warmth, etc.—The best Food for Horses—Work and Digestion—Bulk of Food—Quantity—Water—General Management of the Horse—Air—Litter—Grooming—Exercise—Vices and Habits—Restiveness—Backing and Balking—Biting—Kicking—Running Away—Rearing—Overreaching—Rolling—Shying—Slipping the Halter—Tripping—Hints to Buyers—Warranty—Form of a Receipt Embodying a Warranty—What a Warranty Includes—What constitutes Unsoundness .... Page 9

II.—THE ASS AND THE MULE.

Why the Ass has been Neglected and Abused—Eastern Appreciation—The Ass compared with the Horse—The Ass in Guinea and Persia—The Mule—Adaptation as a Beast of Burden—Trade in Kentucky—Use on a Farm—How to have large and handsome Mules............ 45

III.—CATTLE.

Historical Sketch—Breeds—The Devons—New England Cattle—The Hereford Breed—The Sussex Breed—The Ayrshire Cattle—The Welsh Breeds—Irish Cattle—The Long Horns—The Durham or Short-Horned Breed—Alderney or Jersey Cattle—The Galloway Breed or Hornless Cattle—Other Polled Cattle—The Cream-Pot Breed—Points of Cattle—General Management of Cattle—The Cow-House Feeding—Rearing Calves—Milking—How to Estimate the Weight of Livestock... 49

IV.—SHEEP.

Characteristics of the Sheep—Mutton—Breeds in the United States—The Native Breed—The Spanish Merino—American Merinos—Saxon Merinos—The New Leicester Breed—The South-Downs—Mr. Taylor's Facts and Figures—The Cotswold Breed—New Oxfordshire Sheep—The Cheviot Breed—The

Lincoln Breed—On the Choice of a Breed—The Improved English Varieties as Mutton Sheep—The Merinos as Wool-Producers—General Management—Barns and Sheds—Feeding Racks—Feeding—Salt—Water—Shade—Lambs—Castration—Docking—Washing—Shearing—Value of Sheep to the Farmer—An Anecdote. 73

V.—SWINE.

Natural History of Swine—The Wild Boar—Opinions Respecting the Hog—The Hog among the Greeks and Romans—Swine Breeding in Gaul and Spain—Abhorrence toward Swine's Flesh among the Jews, Egyptians, Mohammedans, and Others—Cuvier's Opinion—Unwholesomeness of Swine's Flesh in Warm Climates—Breeds of Swine—The "Land Pike"—The Chinese Hog—The Berkshire Breed—The Suffolk Breed—The Essex Breed—The Chester Hog—Points of the Hog—Feeding—The Piggery........... 95

VI.—IMPROVEMENT OF BREEDS

Selection of the Sire and Dam—How the Cream Pot Breed was Produced—In-and-In Breeding—Youatt's Opinion—Crossing—Origin of La Chamois Sheep—The best Breeds most Profitable—How to Improve One's Stock—How Improvements may be bred Out as well as In 108

VII.—DISEASES AND THEIR CURE.

About throwing Physic to the Dogs—Wild Animals seldom Sick—The Reason why—Causes of Disease among Domestic Animals—How they may be kept in Perfect Health—Treatment of their Diseases—The Water-Cure for Animals... 114

VIII.—POULTRY.

The Domestic Fowl—Wild Origin Unknown—General Characteristics of the Domestic Fowl—The Spanish Fowl—The Dorking—The Polish Fowl—The Hamburg Fowl—The Dominique Fowl—The Leghorn Fowl—The Shanghais and Cochin Chinas—The Bantam—The Game Fowl—Mongrels—Choice of Breed—Accommodations—Incubation—Rearing Chickens—Five Rules—The Guinea Fowl—The Domestic Turkey—The Principal Requisites in Turkey Rearing—General Directions—The Domestic Goose—How to Rear Geese—Shearing instead of Plucking—The Domestic Duck—Best Varieties—How to Rear Ducks—Fattening—Preparing Poultry for Market 118

IX.—BEE-KEEPING.

Wonders of the Bee-Hive—The three kinds of Bees—The Queen and her Duties—Curious Facts—How the Cells are Made—Bee-Bread—Ventilation by the Bees on Scientific Principles—The Apiary—Bee-Hives—How to Make them—Sectional Hives—Mr. Luda's Hive—Swarming—Robbing the Hive—Wintering—Feeding—Killing the Drones.................................. 148

APPENDIX.

Horse Taming 161

DOMESTIC ANIMALS.

I.

THE HORSE.

A horse! a horse! My kingdom for a horse!—Shakspeare.

I.—HISTORY.

THE horse is probably a native of the warm countries of the East, where he is found wild in a considerable state of perfection. Its use, both as a beast of burden and for the purposes of war, early attracted the attention of mankind. Thus when Joseph proceeded with his father's body from Egypt into Canaan, "there accompanied him both chariots and horsemen" (Gen. xix.); and the Canaanites are said to have gone out to fight against Israel "with many horses and chariots" (Joshua ii. 4). This was more than sixteen hundred years before Christ.

The horse was early employed on the course. In the year 1450 B. C. the Olympic games were established in Greece, at which horses were used in chariot and other races.

No horses were found either on the continent or on the islands of the New World; but the immense droves now existing in parts of both North and South America, all of which have descended from the two or three mares and stallions left by the early Spanish voyagers, prove very clearly that the climate and soil of these countries is well adapted to their propagation.

Professor Low says: "The horse is seen to be affected in his

character and form by the agencies of food and climate, and it may be by other causes unknown to us. He sustains the temperature of the most burning regions; but there is a degree of cold at which he can not exist, and as he approaches this limit his temperament and external conformation are affected. In Iceland, at the Arctic Circle, he has become a dwarf; in Lapland, at latitude 65°, he has given place to the reindeer; and in Kamtschatka, at 62°, he has given place to the dog. The nature and abundance of his food, too, greatly affect his character and form. A country of heaths and innutritious herbs will not produce a horse so large and strong as one of plentiful herbage; the horse of the mountains will be smaller than that of the plains; the horse of the sandy desert than that of the watered valley."*

II.—BREEDS.

The genus *Equus*, according to modern naturalists, consists of six different animals—the horse (*E. caballus*); the ass (*E. asinus*); the quagga (*E. quagga*); the dziggithai (*E. hemionus*); the mountain zebra (*E. zebra*); and the zebra of the plains (*E. burchelli*).

Of the horse there are many varieties or breeds. Ineffectual attempts have been made to decide which variety now existing constitutes the original breed; some contending for the Barb and others for the wild horses of Tartary. It is of the latter that Byron thus speaks in "Mazeppa:"

> With flowing tail and flying mane,
> With nostrils never streaked with pain,
> Mouths bloodless to the bit or rein,
> And feet that iron never shod,
> And flanks unscarred by spur or rod,
> A thousand horse—the wild, the free—
> Likes waves that follow o'er the sea,
> Came thundering on.

The principal breeds of horses now bred in the United States are the Race-Horse, the Arabian, the Morgan, the Canadian,

---

* Illustrations of the Breeds of Animals.

the Norman, the Cleveland Bay, the Conestoga, the Virginia Horse, the Clydesdale, and the Wild or Prairie Horse.

1. *The Race-Horse.*—"There is much dispute," Mr. Youatt says, "with regard to the origin of the *Thorough-bred Horse*. By some he is traced through both sire and dam to Eastern parentage; others believe him to be the native horse, improved and perfected by judicious crossings with the Barb, the Turk, or the Arabian. The Steed Book, which is an authority with every English breeder, traces all the old racers to some Eastern origin; or it traces them until the pedigree is lost in the uncertainty of an early period of breeding.

"Whatever may be the truth as to the origin of the race-horse, the strictest attention has for the last fifty years been paid to pedigree. In the descent of almost every modern racer not the slightest flaw can be discovered."

The racer is generally distinguished, according to the same authority, by his beautiful Arabian head; his fine and finely-set neck; his oblique, lengthened shoulders; his well-bent hinder legs; his ample muscular quarters; his flat legs, rather short from the knee downward, although not always so deep as they should be; and his long and elastic pastern.

The use of thorough-bred and half-bred horses for domestic purposes is becoming common in England. The half-bred horse is not only much handsomer than the common horse, but his speed and power of endurance are infinitely greater.

"The acknowledged superiority of Northern carriage and draught stock," the editor of the New York *Spirit of the Times* says, "is owing almost entirely to the fact that thorough-bred horses have found their way North and East from Long Island and New Jersey, where great numbers are annually disposed of that are unsuited to the course."

For the farm, the pure thorough-bred horse would be nearly useless. He lacks weight and substance to give value and power for draught. For road work the same objections will apply, although not to the same extent, perhaps. The best English road horse is a cross of the thorough-bred and the Cleveland.

2. *The Arabian Horse.*—The genealogy of the Arabian horse, according to Arab account, is known for two thousand years. Many of them have written and attested pedigrees extending more than four hundred years, and, with true Eastern exaggeration, traced by oral tradition from the stud of Solomon. A more careful account is kept of these genealogies than of those of the most ancient family of the proudest Arab chief, and very singular precautions are taken to prevent the possibility of fraud, so far as the written pedigree extends.

The head of the Arabian horse is inimitable. The broadness and squareness of the forehead, the shortness and fineness of the muzzle, the prominence and brilliancy of the eye, the smallness of the ears, and the beautiful course of the veins, are its characteristics. In the formation of the shoulders next to the head, the Arabian is superior to any other breed. The withers are high and the shoulder-blades inclined backward, and so nicely adjusted that in descending a hill the point or edge of the ham never ruffles the skin. The fineness of the legs and the oblique position of the pasterns may seem to lessen his strength; but the leg, although small, is flat and wiry, and its bones uncommonly dense.*

Richardson says: "Often may the traveler in the desert, on entering within the folds of a tent, behold the interesting spectacle of a magnificent courser extended upon the ground, and some half dozen little dark-skinned, naked urchins scrambling across her body, or reclining in sleep, some upon her neck, some on her body, and others pillowed upon her heels; nor do the children ever experience injury from their gentle playmate. She recognizes the family of her friend, her patron, and toward them all the natural sweetness of her disposition leans, even to overflowing."

The Arabian horse Tartar, whose portrait we give on the next page, is thus described in the *New England Farmer:* "This beautiful horse was bred by Asa Pingree, of Topsfield, Mass

---

* Youatt.

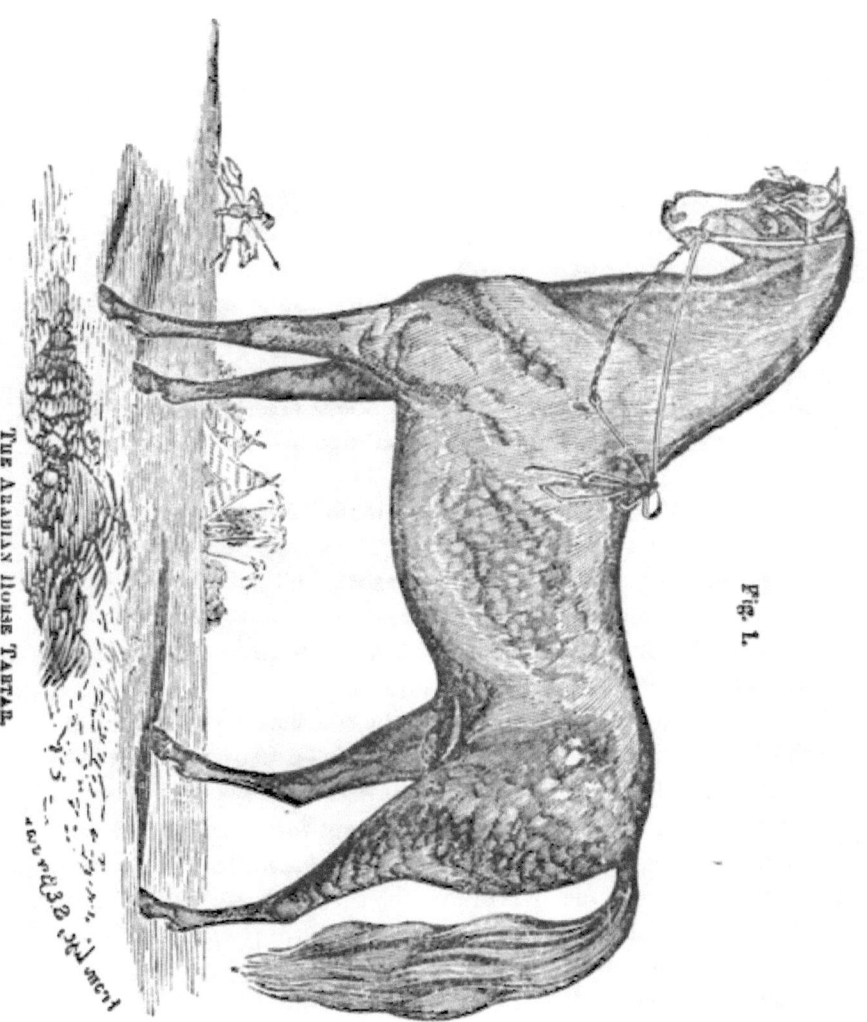

Fig. 1.

THE ARABIAN HORSE TARTAR.

He now stands fifteen and one fourth hands high; weighs nine hundred pounds; is of dark-gray color, with dark mane and tail. He was sired by the imported, full-blood Arabian horse 'Imaum,' and is seven years old this spring. This engraving, copied from life, gives the figure of 'Tartar,' but can not represent the agile action, flashing eye, and cat-like nimbleness of all his movements. It shows the beautiful Arabian head and finely-set-on neck; his ample muscular quarters; his flat legs, rather short from the knee downward; and his long and elastic pastern. All his motions are light and exceedingly graceful, and his temper so docile that a child may handle him."

3. *The Morgan Horse.*—This celebrated American breed is probably a cross between the English race-horse and the common New England mare. It is perhaps, all things considered, the very finest breed for general usefulness now existing in the United States. Mr. S. W. Jewett, a celebrated stock breeder, in an article in the *Cultivator*, says:

"I believe the Morgan blood to be the best ever infused into the Northern horse. The Morgans are well known and esteemed for activity, hardiness, gentleness, and docility; well adapted for all work; good in every spot except for races on the turf. They are lively and spirited, lofty and elegant in their action, carrying themselves gracefully in the harness. They have clean bone, sinewy legs, compactness, short, strong backs, powerful lungs, strength, and endurance. They are known by their short, clean heads, width across the face at the eyes, eyes lively and prominent; they have open and wide under jaws, large windpipe, deep brisket, heavy and round body, broad in the back, short limbs in proportion to size (of body); they have broad quarters, a lively, quick action, indomitable spirit, move true and easy in a good, round trot, and are fast on the walk; color dark bay, chestnut, brown, or black, with dark, flowing, wavy mane and tail. They make the best of roadsters, and live to a great age."

All do not agree, however, with this estimate of the Morgans.

A distinguished judge of horses in Vermont, quoted by Randall in his Introduction to Youatt on the Horse, says:

"They [the Morgans] are good for an hour's drive—for short stages. They are good to run around town with. They are good in the light pleasure-wagon—prompt, lively (not spirited), and 'trappy.' There is no question among those who have had fair opportunities of comparing the Morgans with horses of purer blood and descended from different stocks, in regard to the relative position of the Morgan. He is, as he exists at the present day, inferior in size, speed, and bottom—in fact, in all those qualities necessary to the performance of 'great deeds'

Fig. 2.

SHERMAN MORGAN.

on the road or the farm, to the descendants of Messenger, Duroc, imported Magnum Bonum, and many other horses of deserved celebrity."

Sherman Morgan, whose portrait we are permitted to copy from Linsley's "Morgan Horse," was foaled in 1835, the property of Moses Cook, of Campton, N. H. Sired by Sherman, g sire, Justin Morgan. The pedigree of the dam not fully established, but conceded to have been a very fine animal, and said to

be from the Justin Morgan. Sherman Morgan is fifteen hands high, weighs about 1,050 lbs., is dark chestnut, and very much resembles his sire Sherman, but heavier, stockier, and not as much action. A fine horse, and is now kept in the stable at Lancaster, N. H., where the Sherman died. He is owned by A. J. Congdon.

4. *The Canadian Horse.*—This horse abounds in the Canadian Provinces and in the Northern States of the Union, and is too well known to require a particular description. It is mainly of Norman-French descent. It is a hardy, long-lived animal, is easily kept, and very useful on a farm, although generally too small for heavy work. A cross between stallions of this breed and our common mares produces a superior horse, and such crosses are finding favor among farmers.

5. *The Norman Horse.*—The French or Norman horse, from which the Canadian is descended, is destined to take a more prominent place than has hitherto been assigned to it among our working horses. We introduce an engraving of one of this breed, called Louis Philippe, which was bred by Edward Harris, of Moorestown N. J., by whom the breed was imported from France.

The Norman horse is from the Spanish, of Arabian ancestry, and crossed upon the draught horses of Normandy. Mr. Harris had admired the speed, toughness, and endurance of the French stage-coach horses, and resolved to import this valuable stock, and deserves the thanks of the American public for his perseverance and sacrifices in this enterprise. The Norman horses are enduring and energetic beyond description, and keep their condition on hard fare and brutal treatment, when most other breeds would quail and die. This variety of horse is employed in France to draw the ponderous stage-coaches, called "diligences," and travelers express astonishment at the extraordinary performances of these animals. Each of these huge vehicles is designed for eighteen passengers, and when thus loaded are equal to five tons weight. Five horses are attached to the clumsy and cumbrous carriage, with rude harness, and

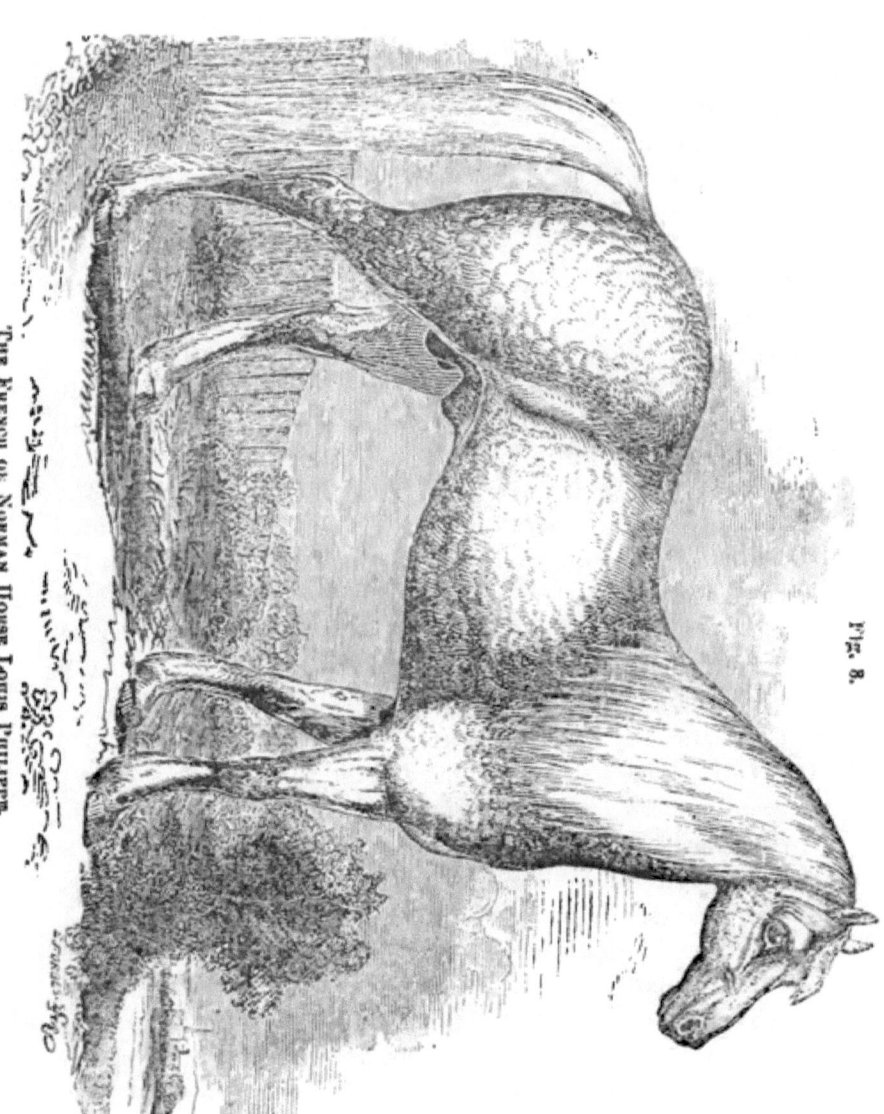

Fig. 8.

The French or Norman Horse Louis Puilippe.

their regular rate of speed with this enormous load is seven miles an hour, and this pace is maintained over rough and hilly regions. On some routes the roads are lighter, when the speed is increased to eight, nine, and sometimes to ten miles an hour.

6. *Cleveland Bay.*—According to Mr. Youatt, the *true* Cleveland Bay is nearly extinct in England. They were formerly employed as a heavy, slow coach-horse. Mr. Youatt says: " The origin of the better kind of coach-horse is the Cleveland Bay, confined principally to Yorkshire and Durham, with perhaps Lincolnshire on one side and Northumberland on the other, but difficult to meet with pure in either county. The Cleveland mare is crossed by a three-fourths or through-bred horse of sufficient substance and height, and the produce is the coach-horse most in repute, with his arched crest and high action. From the thorough-bred of sufficient height, but not of so much substance, we obtain the four-in-hand and superior curricle-horse.

Cleveland Bays were imported into western New York a few years since, where they have spread considerably. They have often been exhibited at our State fairs. They are monstrously large, and for their size are symmetrical horses, and possess very respectable action. Whether they would endure on the road at any but a moderate pace, we are not informed, and have some doubts. Whether they spring from the genuine and unmixed Cleveland stock, now so scarce in England, we have no means of knowing. The half-bloods, the produce of a cross with our common mares, are liked by many of our farmers. They are said to make strong, serviceable farm beasts—though rather prone to sullenness of temper.*

7. *The Conestoga Horse.*—This horse, which is found chiefly in Pennsylvania and the adjacent States, is more remarkable for endurance than symmetry. In height it sometimes reaches seventeen hands; the legs being long and the body light.

---

* Randall.

The Conestoga breed makes good carriage and heavy draft horses.

8. *The Clydesdale Horse.*—The Clydesdale horse is descended from a cross between the Flemish horse and the Lanarkshire (Scotland) mares. The mare is derived from the district on the Clyde where the breed is chiefly found. Horses of this breed are deservedly esteemed for the cart and for the plow on heavy soil. They are strong, hardy, steady, true pullers, of sound constitution, and from fourteen to sixteen hands high. They are broad, thick, heavy, compact, well made for durabil-

Fig. 4.

THE CLYDESDALE HORSE.

ity, health, and power. They have sturdy legs, strong shoulders, back, and hips, a well-arched neck, and a light face and head.

9. *The Virginia Horse.*—This breed predominates in the State from which it takes its name, and abounds to a greater or less extent in all the Southern, Western, and Middle States. It derives its origin from English blood-horses imported at various times, and has been most diligently and purely kept in the South. The celebrated Shark, the best horse of his day,

was sire of the best Virginian horses, while Tally-ho, son of Highflyer, peopled the Jerseys.*

10. *The Wild or Prairie Horse.* — In the Southwestern States wild horses abound, which are doubtless sprung from the same Spanish stock as the wild horses of the pampas and other parts of the southern continent, all of which are of the celebrated Andalusian breed, derived from the Moorish Barb. The prairie horses are often captured, and when domesticated are found to be capable of great endurance. They are not, however, recommended by the symmetry or elegance of appearance for which their type is so greatly distinguished, being generally rather small and scrubby.†

11. *The American Trotting-Horse.*—" We can not refrain," H. S. Randall says, in the Introduction to Youatt on the Horse, already referred to, "from calling attention to our trotting-horses, though in reality they do not, at least as a whole, constitute a breed, or even a distinct variety or family. There *is* a family of superior trotters, including several of the best our country has ever produced, the descendants of Abdallah and Messenger, and running back through their sire Mambrino to the thorough-bred horse, old Messenger. But many of our best trotters have no known pedigrees, and some of them, without doubt, are entirely destitute of the blood of the race-horse. Lady Suffolk is by Engineer, but the blood of Engineer is unknown (she is a gray mare, fifteen hands and two inches high). Dutchman has no known pedigree. Other celebrated trotters stand in the same category—though we are inclined to think that a decided majority of the best, especially at long distances, have a greater or less infusion of the blood of the race-horse.

"The United States has undoubtedly produced more superior trotters than any other country in the world, and in no other country has the speed of the best American trotters been equaled."

---

* Farmers' Register. † Farmers' Encyclopedia.

III.—POINTS OF HORSES.

Every one who has anything to do with the horse should know something of the "points" by means of which a good animal is distinguished from a bad one. It is necessary to understand this, no matter for what particular service the horse may be required; and the qualities indicated by these points are universal in all breeds.

To illustrate this subject and teach the uninstructed how to correctly judge the horse, we introduce the accompanying lettered outlines.

It is evident that to be a good judge of a horse, one must have in his memory a model by which to try all that may be presented to his criticism and judgment.

Fig. 5 represents such a model. It is a thorough-bred horse, in which the artist has endeavored to avoid every fault. Fig. 6 is designed to represent a horse in which every good point is suppressed. It may not be common to see a horse totally destitute of every good point; but injudicious breeding has so obliterated the good ones, that the cut fig. 6 is not a caricature, though we confess that its original is little less than a caricature on the true ideal of a horse. Such a head is common, so is such a shoulder, such a back, quarters, and legs; and if they are not very often all combined in one animal, they are, unfortunately, often found distributed among the common breeds in such abundance as to mar the beauty and the service of three quarters of all the horses in ordinary use. The letters are alike on both figures, and will enable the reader to draw a comparison between the respective points of each. We copy the description of the cuts from the *Farmer's Companion:*

"The most important part of all is probably the direction of the shoulder, from A to B. Next to this, the length from the hip to the hock, C to D. The point which next to these probably most contributes to speed and easy going, is the shortness of the canon bone between the knee and the pastern joint, E to F, a point without which no leg is good. A horse which has all these three points good will necessarily and infallibly

stand over a great deal of ground, W to X, that is, the distance between his fore and hind feet will be great; while one which is deficient in all of them, or, indeed, in the two first, will as

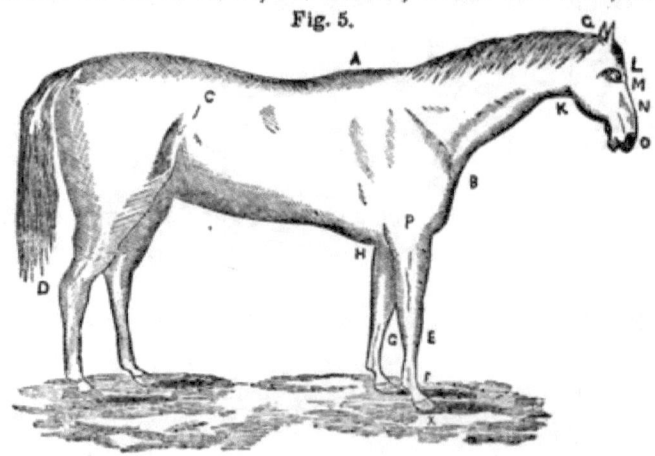

Fig. 5.

assuredly stand like a goat with all its feet gathered under him, and will never be either a fast horse or safe under saddle. A

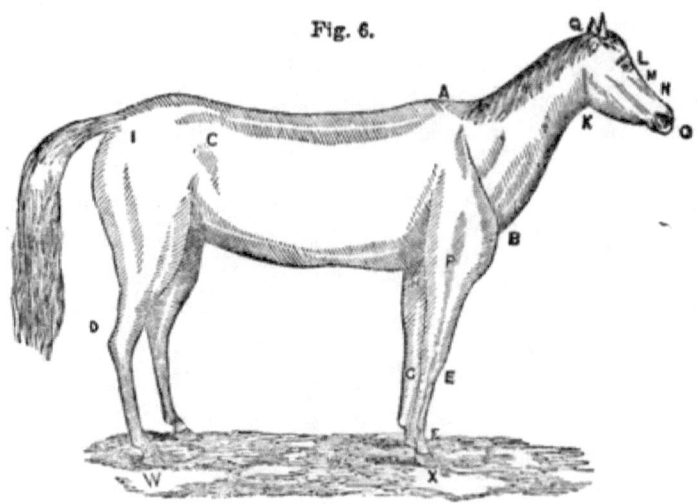

Fig. 6.

horse, not in motion, may be more speedily judged of by this feature than by any other. One consequence of a fine receding shoulder is to give length in the *humerus*, or upper arm, from

B to P, without which a great stride can hardly be attained, but which will seldom if ever be found wanting if the shoulder-blade be well placed. A prominent and fleshy chest is admired by some, probably because they think it indicative of powerful lungs and room for their use. We object to it as adding to what it is so desirable to avoid—the weight to be lifted forward in the act of progression—while all the space the lungs require is to be obtained by *depth* instead of *breadth*, as from A to H, in which point, if a horse be deficient, he will seldom be fit for fast work. The other points which we have marked for comparison are G to E, or the width of the leg immediately below the knee, which in a well-formed leg will be equal all the way down; in a bad one it will be narrowish immediately below the knee, or what is called '*tied in*.' The shape of the neck is more important than might at first thought be supposed, as affecting both the wind and the *handiness* of the mouth; no horse with a faulty neck and a head ill-attached to it, as at Q to K in fig. 6, ever possesses a good or manageable mouth. The points of the face are not without significance, a feebly developed countenance generally showing weakness of courage if not of constitution. We therefore like to see a large and bony protuberance above the eye, as at L in fig. 5, giving the appearance of a sinking immediately below, followed by a slightly *Roman* or protruding inclination toward the nose. These when present are generally signs of 'blood,' which is in some proportion or other a quality without which no breed of horses will ever improve or long entitle itself to rank as other than a race of drudges, fit only for sand or manure carts."

Bearing these points in mind, you may, by observing and comparing the different animals which fall under your eyes, soon qualify yourself to give an intelligent opinion of a horse. One can not become perfect in this branch of knowledge in a week or in a year. Certainly no careful student of this little book will allow himself to be imposed upon in the purchase of an animal having many of the bad points represented in fig. 6. The perfect horse (fig. 5) you will not expect to meet every day.

A badly formed horse is not profitable for any purpose; because, if so formed, they are either clumsy, inactive, dull in mind, or tender and easily broken down. It costs just as much to breed, raise, and keep a poor horse as a good one, and the poor one is low in value and unsalable; besides, he is unable to do good service in any sphere, or to endure.

We copy from Lavater six heads of horses, which indicate different temperaments and a great diversity of character and disposition. The accompanying remarks are from the *American Phrenological Journal:*

Fig. 7.

"Fig. 7 has a slow, heavy temperament; is without spirit, awkward in motion, lazy, stupid in intellect, difficult to teach, bears the whip and needs it, though it is soon forgotten. He is too lazy to hold up his ears or under lip, and is a regular hog-necked, heavy-footed animal.

Fig. 8.

"Fig. 8 has more intelligence and spirit, a more active temperament, and is disposed to anger, will not bear the whip, and shows his anger, when teased or irritated, in a bold, direct onset with the teeth.

Fig. 9.

"Fig. 9 is a very active temperament; is a quick, keen, active, intelligent animal, but is sly, cunning, mischievous, and trickish; will be hard to catch in the field, inclined to slip the bridle, will be a great shirk in double harness, and will require a sharp eye and steady hand to drive him, and will want something besides a frolicsome boy for a master."

"Fig. 10 is obstinate, headstrong, easily irritated, deceitful, and savage; will be hard to drive, unhandy, unyielding, sour-tempered, bad to back, inclined to balk, disposed to fight and crowd his mate, and bite and kick his driver."

Fig. 10.

"Fig. 11 has a noble, proud disposition, and a lofty, stately carriage, but he is timid, restive, and easily irritated and thrown off his mental balance. Such horses should be used by steady, calm men, and on roads and in business which have little variety, change, or means of excitement.

Fig. 11.

Fig. 12.

"Fig. 12 is a calm, self-possessed animal, with a noble, elevated disposition, trustworthy, courageous, good-tempered, well adapted to family use, but not remarkable for sharpness of mind or activity of body.

Fig. 13.

"Figs. 13 and 14 show a great contrast in shape of head, expression of countenance, temperament, disposition, and intelligence. The first is a most noble animal.

"Fig. 13 is broad between the eyes, full, rounded, and prominent in the

2

forehead, indicating benevolence and intellect; broad between the ears, showing courage; broad between the eyes, evincing quickness of perception, memory, and capacity to learn. He can be taught almost anything, can be trusted, and loves and trusts man; is not timid, will go anywhere, and stand without fastening; never kicks, bites, or runs away.

Fig. 14.

"Fig. 14 shows a marked contrast with fig. 13 in almost every respect; his narrow and contracted forehead shows a lack of intelligence, kindness, and tractability; is timid and shy in harness, vicious, unfriendly, disposed to kick, bite, balk, or run away, and is fit only for a mill or horse-boat. For all general uses he should be avoided, and by no means should such an organization be employed for breeding purposes."

IV.—COLOR.

W. C. Spooner, author of several veterinarian works, has the following remarks on color as a sign of other qualities in the horse:

"We have found both good and bad horses of every color, and the only rule we can admit as correct is, that certain colors denote deficient breeding, and therefore such animal is not likely to be so good as he looks, but is probably deficient in bottom or the powers of endurance. These colors are black, which prevails so much with cart-horses, and sorrel, dun, piebald, etc.; the possessors of which come from the North, and possess no Eastern blood. Black horses, unless evidently high bred, are very often soft and sluggish, with breeding insufficient for their work; the pedigree of the majority of them may be dated from the plow-tail, whatever admixtures there may have been since. White hair denotes a thin skin, which

is objectionable when it prevails on the legs of horses, as such

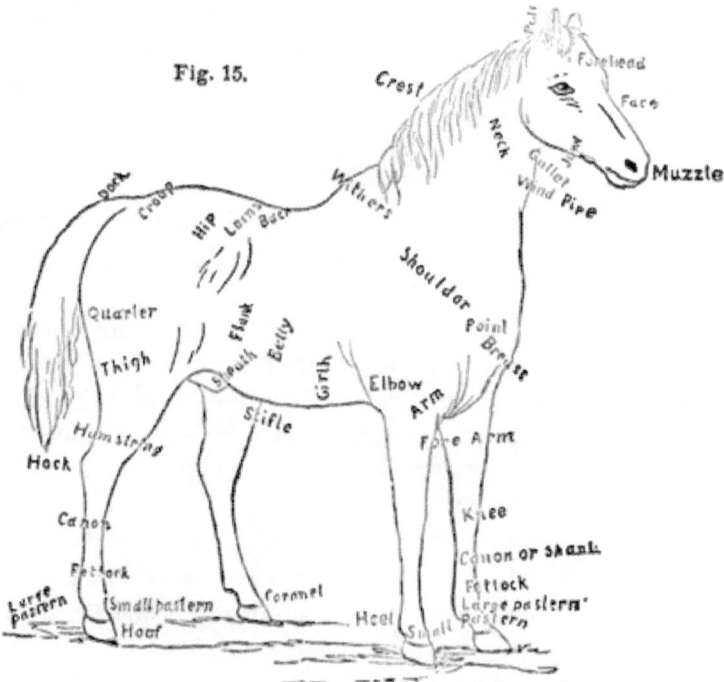

COMMON TERMS DENOTING THE PARTS OF A HORSE.

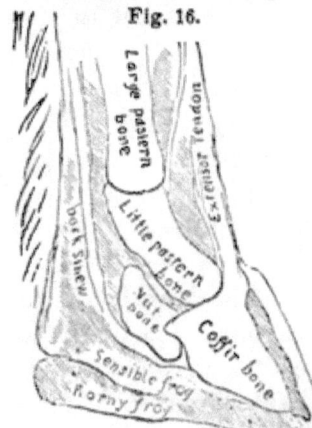

SECTION OF A HORSE'S FOOT.

animals are more disposed to swelled legs and cracked heels than others. Bay horses with black legs are greatly esteemed, yet we have known many determined slugs of this hue. Their constitution is, however, almost invariably good. Chestnut is the prevailing color with our race-horses, and consequently chestnut horses are generally pretty well bred, and possess the good and bad qualities which obtain most among thorough-breds. The Suffolk cart-horse is also distinguished by his light chestnut col-

or; and it is no small recommendation to find that this breed has, for several years past, carried away the principal prizes at the annual shows of the Royal Agricultural Society of England. Gray is a very good color, and generally denotes a considerable admixture of Eastern blood."

V.—STABLES.

We condense from Stewart's admirable "Stable Book" the larger portion of the following useful hints and suggestions in reference to stables and their management.

1. *Stables as they Are.*—Stable architects have not much to boast of. When left to themselves they seem to think of little beyond shelter and confinement. If the weather be kept out and the horse be kept in, the stable is sufficient. If light and air be demanded, the doorway will admit them, and other apertures are superfluous.

The majority of stables have been built with little regard to the comfort and health of the horse. Most of them are too small, too dark, too close, or too open; and some are mere dungeons, destitute of every convenience.

2. *Situation of Stables.*—When any choice exists, a situation should be chosen which admits of draining, shelter from the coldest winds, and facility of access. Damp places are especially to be avoided. It is in damp stables that we expect to find horses with bad eyes, coughs, greasy heels, swelled legs, mange, and a long, dry, staring coat, which no grooming can cure. Take every precaution, then, against dampness in your stables.

3. *Size of Stables.*—They are seldom too large in proportion to the number of stalls; but are often made to hold too many horses. Horses require pure air as well as human beings; and the process of breathing has the same effect in their case as in ours—changing it to that poisonous substance, carbonic acid gas. With twenty or thirty horses in a single apartment no ordinary ventilation is sufficient to keep the air pure. Large stables, too, are liable to frequent and great alterations of temperature. When several horses are out, those which remain

are often rendered uncomfortably cold, and when the stable is full the whole are fevered or excited by excess of heat. Efficient ventilation—a very important object—is also much more difficult in a large than in a small stable.

In width the stable may vary from sixteen to eighteen feet; and in length it must have six feet for each stall. Large carthorses require a little more room both in length and breadth of stable. The number of stalls should not exceed sixteen, and it would be better if there were only eight.

Double-rowed stables, or those in which the stalls occupy both sides, require least space, and for horses kept at full work are sufficiently suitable, but for carriage horses single-rowed stables are better. If the double-rowed are used, the gangway should be wide, to prevent the horses from kicking at each other, as they are apt to do when they grow playful from half idleness.

4. *Windows.*—Windows are too much neglected in stables, and where they exist at all are generally too few, too small, and ill placed. Some think horses do not require light—that they thrive best in the dark; but many a horse has become blind for the want of light in his stable. When side windows can not be introduced, a portion of the hay loft must be sacrificed and light introduced from the roof. Side windows should be so arranged that the light will not fall directly upon the eyes of the horse.

5. *Floors.*—Stable floors may be of stone, brick, plank, or earth. One of the best kinds of stable floor, where the soil is dry, is made of a composition of lime, ashes, and clay, mixed up in equal parts into a mortar and spread from twelve to fifteen inches deep over the surface of the ground forming the bottom of the stable. It will dry in ten days and makes a very smooth, fine flooring, particularly safe, easy, and agreeable for horses to stand upon, and free from all the objections to stone, brick, and wood.*

---

* A. B. Allen.

6. *Draining.*—A gutter or other contrivance for carrying off the urine should always be made in a stable, otherwise it will be foul and damp. It should be conveyed into a tank and carefully saved as manure.

7. *Racks and Mangers.*—These should be so placed that the horse can eat from them with ease. The face of the rack next the horse should be perpendicular, or as nearly so as possible. Sometimes the face is so sloping and the rack so high that the horse has to turn his head almost upside down to get at his food.

The mangers or troughs from which the horse eats his grain are now sometimes made of cast iron, which we deem a great improvement over wood. The manger should be concave and not flat at the bottom. Mangers are generally placed too low. The bottom should be from three feet and a half to four feet from the ground, according to the height of the horse.

8. *Ventilation of Stables.*—Impure air, as we have already remarked, is hurtful to the horse as well as to the human being inducing disease and shortening life. To avoid it in our own case, we (sometimes!) ventilate our houses. If we would have our horses healthy we must do the same for the stable. Apertures, one for each stall, should be provided for carrying off the impure air. These should be so near the top of the building as practicable. It should be eight or ten inches square. Smaller apertures near the floor or not far from the horse's nostrils will serve to admit fresh air.*

9. *Warmth, etc.*—If you wish to have your horses thrive and continue healthy, you can not pay too much attention to their comfort. Their stables should be warm in winter and cool in summer. To secure these conditions, they should be properly constructed. [For plans, see "The House."] To keep stables sufficiently warm, no artificial means are required. It is enough that the outside air, except so much as is required for ventilation, be excluded during the coldest weather. Warm blankets should of course be used at the same time.

---

* See Chapter on Barns and Stables, in "The House."

VI.—FEEDING.

1. *The Best Food for Horses.*—Considerable care and system are necessary in feeding horses, so as to keep them in the best health and the highest working order.

"The best food for ordinary working-horses in America," A. B. Allen says, "is as much good hay or grass as they will eat, corn-stalks or blades, or for the want of these, straw, and a mixture of from sixteen to twenty-four quarts per day, of about half and half of oats and the better quality of wheat bran. When the horse is seven years old past, two to four quarts of corn or hominy or meal ground from the corn and cob is preferable to the pure grain. Two to four quarts of wheat, barley, rye, buckwheat, peas, or beans, either whole or ground, may be substituted for the corn. A pint of oil meal or a gill of flax-seed mixed with the other food is very good for a relish, especially in keeping up a healthy system and the bowels open, and in giving the hair a fine glossy appearance. Potatoes and other roots, unless cooked, do not seem to be of much benefit in this climate, especially in winter—they lie cold upon the stomach and subject the horse to scouring; besides, they are too watery for a hard-working animal. Corn is fed too much at the South and West. It makes horses fat, but can not give them that hard, muscular flesh which oats do; hence their softness and want of endurance in general work and on the road, in comparison with Northern and Eastern horses, reared and fed on oats and more nutritious grasses."

2. *Work and Digestion.*—Slow work aids digestion, empties the bowels, and sharpens the appetite. Hence it happens that on Sunday night and Monday morning there are more cases of colic and founder than during any other part of the week. Horses that never want an appetite ought not to have an unlimited allowance of hay on Sunday; they have time to eat a great deal more than they need, and the torpid state of the stomach and bowels, produced by a day of idleness, renders an additional quantity very dangerous. Farm and cart horses are fed immediately before commencing their labor, and the

appetite with which they return shows that the stomach is **not** full.

During fast work digestion is suspended. In the general commotion excited by violent exertion, the stomach can hardly be in a favorable condition for performing its duty. The blood circulates too rapidly to permit the formation of gastric juice or its combination with the food; and the blood and the nervous influence are so exclusively concentrated and expended upon the muscular system, that none can be spared for carrying on the digestive process.

3. *Bulk Essential.*—Condensed food is necessary for fast-working horses. Their food must be in less compass than that of the farm or cart horse. But to this condensation there are some limits. Grain affords all, and more than all, the nutriment a horse is capable of consuming, even under the most extraordinary exertion. His stomach and bowels can hold more than they are able to digest. Something more than nutriment is wanted. The bowels must suffer a moderate degree of distension; more than a wholesome allowance of grain can produce.

When hay is very dear and grain cheap, it is customary in many stables to give less than the usual allowance of hay, and more grain. The alteration is sometimes carried too far, and is often made too suddenly. The horses may have as much grain as they will eat, yet it does not suffice without fodder. Having no hay, they will leave the grain to eat the litter. When the ordinary fodder, then, is very dear, its place must be supplied by some other which will produce a wholesome distension, though it may not yield so much nutriment. Straw or roots, either or both, may be used in such cases. The excessively tucked-up flank, and the horse's repeated efforts to eat his litter, show when his food is not of sufficient bulk, and this indication must not be disregarded.

4. *Quantity of Food.*—The quantity of food may be insufficient, or it may be in excess. The consumption is influenced by the work, the weather, the horse's condition, age, temper,

form, and health; these circumstances, especially the work, must regulate the allowance.

When the horse has to work as much and as often as he is able, his allowance of food should be unlimited.

When the work is such as to destroy the legs more than it exhausts the system, the food must be given with some restriction, unless the horse be a poor eater.

When the work is moderate, or less than moderate, a good feeder will eat too much.

When the weather is cold, horses that are much exposed to it require more food than when the weather is warm.

When the horse is in good working condition, he needs less food than while he is only getting into condition.

Young growing horses require a little more food than those of mature age; but, as they are not fit for full work, the difference is not great.

Old horses, those that have begun to decline in vigor, require more food than the young or the matured.

Hot-tempered, irritable horses seldom feed well; but those that have good appetites require more food to keep them in condition, than others of quiet and calm disposition.

Small-bellied, narrow-chested horses require more food than those of deep and round carcass; but few of them eat enough to maintain them in condition for full work.

Lame, greasy-heeled, and harness-galled horses require an extra allowance of food to keep them in working condition.

Sickness, fevers, inflammations, all diseases which influence health so much as to throw the horse off work, demand, with few exceptions, a spare diet, which, in general, consists of bran-mashes, grass, carrots, and hay.*

5. *Watering.*—This is a part of stable management little regarded by the farmer. He lets his horses loose morning and night, and they go to the nearest pond or brook and drink their fill, and no harm results, for they obtain that kind of water

---

* Stewart's Stable Book.

which nature designed them to have, in a manner prepared for them by some unknown influence of the atmosphere, as well as by the deposition of many saline admixtures.

The difference between *hard* and *soft* water is known to every one. There is nothing in which the different effect of hard and soft water is so evident as in the stomach and digestive organs of the horse. Hard water drawn fresh from the well will assuredly make the coat of a horse unaccustomed to it stare, and it will not unfrequently gripe and otherwise injure him. He is injured, however, not so much by the hardness of the well-water as by its coldness—particularly by its coldness in summer, and when it is in many degrees below the temperature of the atmosphere. The water in the brook and the pond being warmed by long exposure to the air, as well as having become soft, the horse drinks freely of it without danger.

If the horse were watered three times a day, and especially in summer, he would often be saved from the sad torture of thirst, and from many a disease. Whoever has observed the eagerness with which the over-worked horse, hot and tired, plunges his muzzle into the pail, and the difficulty of stopping him until he has drained the last drop, may form some idea of what he had previously suffered, and will not wonder at the violent spasms, and inflammation, and sudden death that often result. There is a prejudice in the minds of many persons against the horse being fully supplied with water. They think that it injures his wind, and disables him for quick and hard work. If he is galloped, as he too often is, immediately after drinking, his wind may be irreparably injured; but if he were oftener suffered to satiate his thirst at the intervals of rest he would be happier and better. It is a fact unsuspected by those who have not carefully observed the horse, that if he has frequent access to water he will not drink so much in the course of the day as another will do, who, to cool his parched mouth, swallows as fast as he can and knows not when to stop.

On a journey a horse should be liberally supplied with

water. When he is a little cooled, two or three quarts may be given to him, and after that his feed. Before he has finished his corn, two or three quarts more may be offered. He will take no harm if this is repeated three or four times during a long and hot day.*

VII.—GENERAL MANAGEMENT.

1. *Air.*—We have spoken of the necessity of ventilation. Hear what that great authority, Youatt, says:

"If the stable is close, the air will not only be hot but foul. The breathing of every animal contaminates it; and when in the course of the night, with every aperture stopped, it passes again and again through the lungs, the blood can not undergo its proper and healthy change; digestion will not be so perfectly performed, and all the functions of life are injured. Let the owner of a valuable horse think of his passing twenty or twenty-two out of the twenty-four hours in this debilitating atmosphere! Nature does wonders in enabling every animal to accommodate itself to the situation in which it is placed, and the horse that lives in the stable-oven suffers less from it than would scarcely be conceived possible: but he does not, and can not, possess the power and the hardihood which he would acquire under other circumstances.

"The air of the improperly close and heated stable is still further contaminated by the urine and dung, which rapidly ferment there, and give out stimulating and unwholesome vapors. When a person first enters an ill-managed stable, and especially early in the morning, he is annoyed not only by the heat of the confined air, but by a pungent smell, resembling hartshorn; and can he be surprised at the inflammation of the eyes, and the chronic cough, and the disease of the lungs, by which the animal, who has been all night shut up in this vitiated atmosphere, is often attacked; or if glanders and farcy should occasionally break out in such stables? It has been ascertained by chemical experiment that the urine of the horse

---

* Youatt.

contains in it an exceedingly large quantity of hartshorn; and not only so, but that, influenced by the heat of a crowded stable, and possibly by other decompositions that are going forward at the same time, this ammoniacal vapor begins to be rapidly given out almost immediately after the urine is voided."

2. *Litter.*—The facts just stated in reference to the plentiful escape of ammoniacal gas from the urine, show the necessity of frequently removing the litter which is soon saturated with it. It rapidly putrefies, emitting noisome odors and contaminating the air. Everything hastening decomposition should be carefully removed where life and health are to be preserved. Litter that has been much wetted and has begun to decay should be swept out every morning.

No heap of fermenting dung should be suffered to remain during the day in the corner or any part of the stable.

3. *Grooming.*—Of this little need be said to the farmer in reference to his working horses, since custom, and apparently without ill effect, has allotted to them so little of the comb and brush. "The animal that is worked all day and turned out at night," Youatt says, "requires little more to be done to him than to have the dirt brushed off his limbs. Regular grooming, by rendering his skin more sensible to the alteration of temperature and the inclemency of the weather, would be prejudicial. The horse that is altogether turned out, needs no grooming. The dandruff or scurf, which accumulates at the roots of the hair, is a provision of nature to defend him from the wind and the cold.

"It is to the stabled horse, highly fed and little or irregularly worked, that grooming is of so much consequence. Good rubbing with the brush or the curry-comb opens the pores of the skin, circulates the blood to the extremities of the body, produces free and healthy perspiration, and stands in the room of exercise. No horse will carry a fine coat without either unnatural heat or dressing. They both effect the same purpose; they both increase the insensible perspiration; but the first does it at the expense of health and strength, while

the second, at the same time that it produces a glow on the skin and a determination of blood to it, rouses all the energies of the frame. It would be well for the proprietor of the horse if he were to insist—and to see that his orders are really obeyed—that the fine coat in which he and his groom so much delight is produced by honest rubbing, and not by a heated stable and thick clothing, and, most of all, not by stimulating or injurious spices. The horse should be regularly dressed every day, in addition to the grooming that is necessary after work.

"When the weather will permit the horse to be taken out, he should never be groomed in the stable, unless he is an animal of peculiar value, or placed for a time under peculiar circumstances. Without dwelling on the want of cleanliness, when the scurf and dust that are brushed from the horse lodge in his manger and mingle with his food, experience teaches, that if the cold is not too great, the animal is braced and invigorated to a degree that can not be attained in the stable, from being dressed in the open air. There is no necessity, however, for half the punishment which many a groom inflicts upon the horse in the act of dressing; and particularly on one whose skin is thin and sensible. The curry-comb should at all times be lightly applied. With many horses, its use may be almost dispensed with; and even the brush needs not to be so hard, nor the points of the bristles so irregular, as they often are. A soft brush, with a little more weight of the hand, will be equally effectual and a great deal more pleasant to the horse. A hair-cloth, while it will seldom irritate and tease, will be almost sufficient with horses that have a thin skin, and that have not been neglected. After all, it is no slight task to dress a horse as it ought to be done. It occupies no little time, and demands considerable patience as well as dexterity. It will be readily ascertained whether a horse has been well dressed, by rubbing him with one of the fingers. A greasy stain will detect the idleness of the groom. When, however, the horse is changing his coat, both the curry-comb and the brush should be used as lightly as possible.

"Whoever would be convinced of the benefit of friction to the horse's skin and to the horse generally, needs only to observe the effects produced by well hand-rubbing the legs of a tired horse. While every enlargement subsides, and the painful stiffness disappears, and the legs attain their natural warmth and become fine, the animal is evidently and rapidly reviving; he attacks his food with appetite, and then quietly lies down to rest."

4. *Exercise.*—Of this the farm horse generally has enough. His work is tolerably regular, not exhausting, and he generally maintains his health and has his life prolonged to an extent rare among horses of "leisure." But a gentleman's or a tradesman's horse suffers a great deal more from idleness than he does from work. A stable-fed horse should have two hours' exercise every day, if he is to be kept free from disease. Nothing of extraordinary, or even of ordinary, labor can be effected on the road or in the field without sufficient and regular exercise. It is this alone which can give energy to the system or develope the powers of any animal. The animal that, with the usual stable feeding, stands idle for three or four days, as is the case in many establishments, must suffer. He is predisposed to fever, or to grease, or, most of all, to diseases of the foot; and if, after three or four days of inactivity he is ridden far and fast he is almost sure to have inflammation of the lungs or of the feet.

VIII.—VICES AND BAD HABITS.

The vices and bad habits of the horse, like those of his master, are oftener than otherwise the consequence of a faulty education. We are convinced that innately vicious horses are comparatively few. We condense from Youatt the following hints on this subject.

1. *Restiveness.*—At the head of all the vices of the horse is restiveness, the most annoying and the most dangerous of all. It is the produce of bad temper and worse education; and, like all other habits founded on nature and stamped by education, it is inveterate. Whether it appears in the form of

kicking or rearing, plunging or bolting, or in any way that threatens danger to the rider or the horse, it rarely admits of cure. A determined rider may, to a certain extent, subjugate the animal; or the horse may have his favorites, or form his attachments, and with some particular person he may be comparatively or perfectly manageable; but others can not long depend upon him, and even his master is not always sure of him.

2. *Backing or Balking.*—Some horses have the habit of backing at first starting, and that more from playfulness than desire of mischief. A moderate application of the whip will usually be effectual. Others, even after starting, exhibit considerable obstinacy and viciousness. This is frequently the effect of bad breaking.

A large and heavy stone should be put behind the wheel before starting, when the horse, finding it more difficult to back than to go forward, will gradually forget this unpleasant trick. It will likewise be of advantage as often as it can be managed, so to start that the horse shall have to back up-hill. The difficulty of accomplishing this will soon make him readily go forward. A little coaxing or leading will assist in accomplishing the cure.

3. *Biting.*—This is perhaps sometimes the consequence of natural ferocity, but is more frequently acquired from the foolish teasing play of hostlers and stable-boys. At first his biting is half playful and half in earnest, but finally becomes habitual and degenerates into absolute viciousness. It is seldom that anything can be done in the way of cure. Kindness will aggravate the evil and no degree of severity will correct it. "I have seen," Professor Stuart says, "biters punished until they trembled in every joint and were ready to drop, but have never in any case known them cured by this treatment or by any other. The lash is forgotten in an hour, and the horse is as ready and determined to repeat the offense as before. He appears unable to resist the temptation, and in its worst form biting is a species of insanity."

But if biting can not be cured it may almost always be **prevented**, and every proprietor of horses, while he insists upon gentle and humane treatment of his animals, should strictly forbid this horse-play.

4. *Kicking.*—This, as *a vice*, is another consequence of the culpable habit of grooms and stable-boys of teasing the horse. That which is at first an indication of annoyance at the pinching and tickling of the groom, and without any design to injure, gradually becomes the expression of anger, and the effort to do mischief. The horse, likewise, too soon recognizes the least appearance of timidity, and takes advantage of the discovery. There is no cure for this vice after it has become a confirmed habit, and he can not be justified who keeps a kicking horse in his stable. Before the habit is inveterately established, a thorn-bush or a piece of furze fastened against the partition or post will sometimes effect a cure. When the horse finds that he is pretty severely pricked he will not long continue to punish himself.

5. *Rearing.*—This sometimes results from playfulness, carried, indeed, to an unpleasant and dangerous extent; but it is oftener a desperate and occasionally successful effort to unhorse the rider, and consequently a vice. The horse that has twice decidedly and dangerously reared should never be trusted again, unless, indeed, it was the fault of the rider, who had been using a deep curb and a sharp bit. Some of the best horses will contend against these, and then rearing may be immediately and permanently cured by using a snaffle bridle alone.

6. *Running Away.*—There is no certainty of cure for this vice. The only method which affords any probability of success is, to ride or drive such a horse with a strong curb and sharp bit; to have him always firmly in hand; and if he will run away and the place will admit of it, to give him (sparing neither curb nor whip) a great deal more running than he likes.

7. *Overreaching.*—This unpleasant noise, known also by the term "clicking," arises from the toe of the hind-foot knocking

against the shoe of the fore-foot. If the animal is young, the action of the horse may be materially improved; otherwise nothing can be done, except to keep the toe of the hind-foot as short and as round as it can safely be, and to bevel off and round the toe of the shoe, like that which has been worn off by a stumbling horse, and, perhaps, to lower the heel of the fore-foot a little.

8. *Rolling.*—Some horses have the habit of rolling in the stable, by which they are liable to get cast, bruised, and half strangled. The only remedy is to tie such a horse with just length of halter enough to lie down, but not allow of his resting his head on the ground. This is an unpleasant means of cure, and not always a safe one.

9. *Shying.*—This vice is often the result of cowardice, or playfulness, or want of work, but at other times it is the consequence of a defect of sight; and in its treatment it is of great importance to distinguish between these different causes. For the last, every allowance must be made, and care must be taken that the fear of correction is not associated with the imagined existence of some terrifying object. The severe use of the whip and the spur can not do good here, and are likely to aggravate the vice ten-fold. A word half encouraging and half scolding will tell the horse that there was nothing to fear, and will give him confidence in his rider on a future occasion.

The shying from skittishness or affectation is quite a different affair, and must be conquered: but how? Severity is altogether out of place. If he is forced into contact with the object by dint of correction, the dread of punishment will afterward be associated with that object, and on the next occasion his startings will be more frequent and more dangerous. The way to cure him is to go on, turning as little as possible out of the road, giving a harsh word or two and a gentle touch, and then taking no more notice of the matter. After a few times, whatever may have been the object which he chose to select as the pretended cause of affright, he will pass it almost without notice.

10. *Slipping the Halter.*—Many horses are so clever at this

trick that scarcely a night passes without their getting loose. It is a habit which may lead to dangerous results, and should be cured at once by some extra means of securing the halter in its place, or by a strap attached to it and buckled securely (but not tight enough to be a serious inconvenience), around the neck.

11. *Tripping.*—He must be a skillful practitioner or a mere pretender who promises to remedy this habit. If it arises from a heavy fore-hand and the fore-legs being too much under the horse, no one can alter the natural frame of the animal; if it proceeds from tenderness of the foot, grogginess, or old lameness, these ailments are seldom cured. Also, if it is to be traced to habitual carelessness and idleness, no whipping will rouse the drone. A known stumbler should never be ridden or driven by any one who values his safety or his life.

If the stumbler has the foot kept as short and the toe pared as close as safety will permit, and the shoe is rounded at the toe, or has that shape given to it which it naturally acquires in a fortnight from the peculiar action of such a horse, the animal may not stumble quite so much; or if the disease which produced the habit can be alleviated, some trifling good may be done, but in almost every case a stumbler should be got rid of, or put to slow and heavy work. If the latter alternative is adopted, he may trip as much as he pleases, for the weight of the load and the motion of the other horses will keep him upon his legs.

IX.—HINTS TO BUYERS.

1. *Warranty.*—A man should have a more perfect knowledge of horses than falls to the lot of most men, and a perfect knowledge of the seller also, who ventures to buy a horse without a warranty. This warranty is usually embodied in the receipt, which may be expressed as follows:

Received at Louisville, August 10th, 1858, from C. D., one hundred dollars for a gray horse warranted only five years old, sound, free from vice, and quiet to ride or drive. A. V.

"A receipt, including merely the word 'warranted,' ex-

tends only to soundness; 'warranted sound' goes no further; the age, freedom from vice, and quietness to ride and drive, should be especially named. This warranty comprises every cause of unsoundness that can be detected, or that lurks in the constitution at the time of sale, and to every vicious habit that the animal has hitherto shown. To establish a breach of warranty, and to be enabled to tender a return of the horse and recover the difference of price, the purchaser must prove that it was unsound or viciously disposed at the time of sale.

"No price will imply a warranty or be equivalent to one; there must be an express warranty. A fraud must be proved in the seller, in order that the buyer may be enabled to return the horse or maintain an action for the price. The warranty should be given at the time of sale. A warranty, or a promise to warrant the horse, given at any period antecedent to the sale, is invalid; for horse flesh is a very perishable commodity, and the constitution and usefulness of the animal may undergo a considerable change in the space of a few days. A warranty after the sale is invalid, for it is given without any legal consideration. In order to complete the purchase, there must be a transfer of the animal, or a memorandum of agreement, or the payment of the earnest-money. The least sum will suffice for earnest. No verbal promise to buy or to sell is binding without one of these. The moment either of these is effected, the legal transfer of property or delivery is made, and whatever may happen to the horse, the seller retains, or is entitled to, the money. If the purchaser exercises any act of ownership, by using the animal without leave of the vender, or by having any operation performed, or any medicine given to him, he makes him his own.

"If a person buys a horse warranted sound, and discovering no defect in him, and relying on the warranty, re-sells him, and the unsoundness is discovered by the second purchaser, and the horse returned to the first purchaser, or an action commenced against him, he has his claim on the first seller, and may demand of him not only the price of the horse, or the dif-

ference in value, but every expense that may have been incurred.

"Absolute exchanges of one horse for another, or a sum of money being paid in addition by one of the parties, stand on the same ground as simple sales. If there is a warranty on either side, and that is broken, an action may be maintained: if there be no warranty, deceit must be proved."

2. *What constitutes Unsoundness?*—"That horse is sound in whom there is no disease, and no alteration of structure that impairs or is likely to impair his natural usefulness. The horse is unsound that labors under disease, or has some alteration of structure which does interfere, or is likely to interfere, with his natural usefulness. The term '*natural usefulness*' must be borne in mind. One horse may possess great speed, but is soon knocked up; another will work all day, but can not be got beyond a snail's pace; a third with a heavy fore-hand is liable to stumble, and is continually putting to hazard the neck of his rider; another, with an irritable constitution and a loose, washy form, loses his appetite and begins to scour if a little extra work is exacted from him. The term unsoundness must not be applied to either of these; it would be opening far too widely a door to disputation and endless wrangling. The buyer can discern, or ought to know, whether the form of the horse is that which will render him likely to suit his purpose, and he should try him sufficiently to ascertain his natural strength, endurance, and manner of going. Unsoundness, we repeat, has reference only to disease, or to that alteration of structure which is connected with, or will produce, disease and lessen the usefulness of the animal."*

---

* Youatt.

II.

THE ASS AND THE MULE.

<small>O, that I had been writ down an ass!—*Dogberry.*</small>

I.—THE ASS.

BUFFON has well observed that the ass is despised and neglected only because we possess a more noble and powerful animal in the horse, and that if the horse were unknown, and the care and attention that we lavish upon him were transferred to his now neglected and despised rival, the latter would be increased in size and developed in mental qualities to an extent which it would be difficult to anticipate, but which Eastern travelers, who have observed both animals in their native climates, and among nations by whom they are equally valued, and the good qualities of each justly appreciated, assure us to be the fact.

The character and habits of the horse and the ass are in many respects directly opposed. The one is proud, fiery, impetuous, nice in his tastes, and delicate in his constitution; subject, like a pampered menial, to many diseases, and having many wants and habits unknown in a state of nature. The other, on the contrary, is humble, patient, quiet, and hardy.

For food the ass contents himself with the most harsh and disagreeable herbs, which other animals will scarcely touch; in the choice of water he is, however, very nice, drinking only that which is perfectly clear, and at brooks with which he is acquainted.

The qualities of the ass as a working animal are almost or quite unknown in this country, but in other lands he is found

very serviceable to the poor who are not able to buy or to keep horses. He requires very little care, bears correction with firmness, sustains labor and hunger with patience, and is seldom or never sick.

The varieties of the ass, in countries favorable to their development, are great. In Guinea the asses are large, and in shape even excel the native horses. The asses of Arabia (Chardin says) are perhaps the handsomest animals in the world. Their coat is smooth and clean; they carry the head elevated; and have fine and well-formed legs, which they throw out gracefully in walking or galloping. In Persia also they are finely formed, some being even stately, and much used in draught and for carrying burdens, while others are more lightly proportioned, and used for the saddle by persons of quality; frequently fetching the large sum of 400 livres; and being taught a kind of ambling pace, are richly caparisoned and used by the rich and luxurious nobles.*

II.—THE MULE.

The principal objection to the ass, as a beast of burden, being his small size, the ingenuity of man early devised means to remedy this defect by crossing him with the horse; thus producing an intermediate animal with the size and strength of the latter, and the patience, hardiness, and sure-footedness of the former.

The mule is the offspring of the ass and the mare, or the female ass and the horse. In the latter case the produce is called a jennet, and is much less hardy, and therefore rarely bred.

Mules are much used in warm climates, where they are preferred to horses for many purposes. They are very numerous in our Southern States and not uncommon in the Middle and Western States.

Kentucky is the great mule-breeding State. Many thou-

---

* Blane's Encyclopedia of Rural Sports.

sands are annually raised there for the New York and Southern markets. A correspondent of the *American Veterinary Journal* says:

"The mule trade is one of the largest of Kentucky, and affords one of her chief sources of revenue. The mule is fed from weaning time (which is generally at the age of five or six months) to the full extent of its capacity to eat, and that, too, on oats and corn, together with hay and fodder. In lieu of the long food, soiling is usually adopted in the summer, as they are kept confined in a pound or paddock, containing an acre or two of ground, which is usually partially shaded, in herds of one hundred or one hundred and fifty. In this way they are kept until the fall after they are two years old, receiving a sort of forcing hot-house treatment. At this age they are taken to the Southern market, not always by the feeder, but more generally by the speculator or trader; there they are sold to the planter entirely unbroken. The planters are too cautious to buy a broken mule, lest it should prove to be an antiquated, broken-down beast, fattened up and sold for a young one—as it is more difficult to judge of his age than that of a horse. The external marks of time and service are not generally so apparent upon him. But it is a small job to break a mule. It is only necessary to have a steady horse to work him, with a second hand to drive him an hour or two to keep him up, after which he is considered ready for any service that the farmer may require of him. He may kick once or twice, but is unlike the spirited horse, who when he commences is apt to kick himself out of the harness before he stops.

"Persons who have tried them on the farm are pleased with them. They never get sick and rarely get lame, will do as much work as horses which will cost twice as much money, and at the same time will subsist on less and inferior food; for a mule will work very well on wheat straw and corn shucks, whereas the horse must have grain as well as a good allowance of long food. They are better for our servants to handle, as they can stand neglect and violent treatment better than the

horse, and a blemish, such as the loss of an eye, **does not** impair their value so much as that of the horse."

To have large and handsome mules, the mare should be of a large breed and well proportioned, with rather small limbs, a moderate sized head, and a good forehead; and the ass should be of the large Spanish breed.

III.

CATTLE.

*The noble, patient ox and gentle cow
Kind usage claim ; and he's a brute indeed,
Unworthy of companionship with them,
Who with neglect or cruelty repays
The debt he owes their race.—Knox.*

I.—HISTORY

OF the ox tribe (*Bovidæ*) there are eight species—the ancient bison (*Bos urus*); the bison or American buffalo (*B. bison*); the musk ox (*B. moschatus*); the gayal (*B. frontalis*); the grunting ox (*B. grunniens*); the buffalo of Southern Africa (*B. caffer*); the common buffalo (*B. bubulus*); and the common domestic ox (*B. taurus*). It is with the last only that we have to do in the present work.

The ox has been domesticated and in the service of man from the remotest antiquity. The Bible informs us that cattle were kept by the early descendants of Adam (Gen. iv. 20). That their value has been duly appreciated in all ages and in all climates, is shown by authentic history. Both the Hindoos and the Egyptians placed the ox among their deities; and no quadruped certainly is more worthy to be thus exalted.

The parent race of the ox is supposed by some to have been much larger than any of the present varieties. The urus, in his wild state at least, was an enormous and fierce animal, and ancient legends have thrown around him an air of mystery. In almost every part of the continent of Europe and in England, skulls, evidently belonging to cattle, have been found far exceeding in size those of the present day ; but these may have belonged to exceptional individuals.

Of the original race of British cattle no satisfactory description occurs in any ancient author; but it is believed that, with occasional exceptions, they possessed no great bulk or beauty. They were doubtless numerous, for Cæsar tells us, in his Commentaries, that the ancient Britons neglected tillage and lived on milk and flesh. It was that occupation and mode of life which suited their state of society. A few specimens of the pure ancient breed, descendants of cattle which escaped from their masters centuries since and became wild, may now be seen in the parks of gentlemen in England. They are very wild, and are said to be untamable.

The breeds of cattle in England are remarkable for their numerous varieties, caused by the almost endless crossings of one breed with another.

The breeds of cattle now found in America are all derived from Europe, and those of the United States mainly from England. The early importations were of inferior grades, as the grand improvements in British cattle, commenced by Bakewell, date back no farther than about the time of the Revolution. In New England the primitive stock is believed to have undergone considerable improvement, while in parts of the Middle and Southern States it has undoubtedly deteriorated.

II.—BREEDS.

A strict classification of the numerous breeds of cattle now existing in the United States would be difficult. Youatt arranges British cattle under three heads, according to the comparative size of their horns—the Long Horns, the Short Horns, and the Middle Horns. These classes are all represented here. The prevailing stock of the Eastern States is believed to be derived from the Middle Horns or North Devons, most of the excellent marks and qualities of which they possess. They have frequently been called the American Devons, and are highly esteemed. The most valuable working oxen are of this breed, which also contributes largely to the best displays of beef found in the markets of New York, Philadelphia, and Bos-

ton. The Long Horns or Craven cattle, although not numerous, are occasionally met with. The Short Horns are of more recent introduction, but this breed, with various crosses, is now perhaps the predominant one of the country.

It will be profitable to speak somewhat in detail, although briefly, of the several breeds—at least the more prominent ones—and we will begin with

Fig. 17.

A DEVON BULL.

1. *The Devon Breed.*—This is a handsome and valuable breed. The bull should have yellow horns; clear, bright, and prominent eyes; small, flat, indented forehead; a fine muzzle; small cheek; a clear yellow nose; a high and open nostril; a thick neck, with the hair about the head curled; a straight back; and be well set upon the legs. The head of the ox is smaller, otherwise he does not differ materially in shape from

the bull. He is quicker in his motions than any other ox, and is generally docile, good tempered, and honest.

The cow is much smaller than the bull, but roomy for breeding, and distinguished for her clear, round eye and general beauty of features. With regard to the comparative value of the Devon cows for the dairy there is much difference of opinion, it being pretty generally asserted that their acknowledged grazing qualities render them unfit for the dairy, and that their milk is rich but deficient in quantity. Many superior judges,

A DEVON HEIFER.

however, prefer them even for the dairy. Both cows and oxen fatten faster and with less food than most others.* In color Devon cattle are generally red.

Our New England cattle, as we have said, are generally derived from this breed. Their horns are moderately long,

---

* Youatt.

smooth, and slender, and their prevailing color deep red; but sometimes they are dark brown, brindle, or nearly black. The oxen are remarkable for their docility, strength, and quickness. The cows are fair milkers. Both oxen and cows fatten readily.

2. *The Hereford Breed.*—Cattle of the Hereford breed are larger than those of the North Devon. They are broad across the hind-quarters; narrow at the sirloin; neck and head well

Fig. 19.

THE HEREFORD BULL, TROMP.

proportioned; horns of a medium size and turned up at the points; color a deep red, with the face, throat, and belly generally white. A spirited contest has been kept up for some time between the partisans of the Herefords and those of the Short Horns, both here and in England, each stoutly maintaining the superiority of their favorite breed. We are not disposed to take part in the controversy. The experience of persons not

engaged in breeding either sort as a special business must finally settle it; in the mean time, candid people will acknowledge that both are excellent, each in its way.

Youatt says that the Herefords fatten to a much greater weight than the Devons, and that a Hereford cow will grow fat where a Devon would starve. They are very hardy, and will do well with only the same care required by our native breeds.

3. *The Sussex Breed.*—The Sussex ox holds an intermediate place between the Devon and the Hereford; with much of the activity of the first and the strength of the second, and the propensity to fatten, and the beautiful fine-grained flesh of both. Experience has shown that it combines as many of the good qualities of both as can be combined in one frame. The Sussex cow does not answer for the dairy, her milk, although of good quality, is so small in quantity that she is little regarded for making butter and cheese. The prevailing color of the Sussex cattle is a deep chestnut red.*

4. *Ayrshire Breed.*—The Ayrshire breed, which is considered the most valuable in Scotland, is of the small size and middle-horned race. In modern times it has been much improved. Mr. Aiton, in his Survey of Ayrshire, thus describes this fine breed:

"The most approved shapes in the dairy breed are, small head, rather long and narrow at the muzzle; eye small, but smart and lively; the horns small, clear, crooked, and their roots at considerable distance from each other; neck long and slender, tapering toward the head, with no loose skin below; shoulders thin; fore-quarters light; hind-quarters large; back straight, broad behind; the joints rather loose and open; carcass deep, and pelvis capacious and wide over the hips, with round, fleshy buttocks; tail long and small; legs small and short, with firm joints; udder capacious, broad, and square, stretching forward, and neither fleshy, low hung, nor loose; the milk-veins are large and prominent; teats short, all pointing outward, and at considerable distance from each other;

---

* Youatt.

skin thin and loose; hair soft and woolly; the head, bones, horns, and all parts of least value, small; and the general figure compact and well proportioned."

"The qualities of a cow," adds Mr. Aiton in another place, "are of great importance. Tameness and docility of temper greatly enhance the value of a milch cow. Some degree of hardiness, a sound constitution, health, and a moderate degree of spirits, are qualities to be wished for in a dairy cow, and what those of Ayrshire generally possess. The most valuable qualities which a dairy cow can possess are that she yields much milk, and that of an oily, butyraceous, and caseous nature; and that after she has yielded very large quantities of milk for several years, she shall be as valuable for beef as any other breed of cows known; her fat shall be much more mixed through the whole flesh, and she shall fatten faster than any other."

There have been several importations of Ayrshires into the United States, but they have, up to the present time, failed to establish themselves in general favor.

5. *Welsh Cattle.*—"The cattle of Wales are principally of the Middle Horns, and stunted in their growth from the poverty of their pastures. Of these there are several varieties. The Pembrokeshire are chiefly black, with white horns; are shorter legged than most other Welsh cattle; are larger than those of Montgomery, and have round and deep carcasses; have a lively look and good eyes; though short and rough, not thick; have not large bones, and possess, perhaps, as much as possible, the opposite qualities of being very fair milkers, with a propensity to fatten. The meat is equal to the Scotch. They will thrive, says Mr. Youatt, where others starve, and they rapidly outstrip most others when they have plenty of good pasture. The Pembroke cow has been called the poor man's cow. The Pembroke ox is a speedy and an honest worker, and when taken from hard work fattens speedily. Many are brought to London, and rarely disappoint the butcher."

6. *Irish Cattle.*—Of the Irish cattle there are two breeds—

the Middle Horns and the Long Horns. The Middle Horns are the original breed. "They are," Mr. Youatt says, "small, light, active, and wild; the head commonly small; the horns short but fine, rather upright, and frequently, after projecting forward, turning backward; somewhat deficient in hind-quarters; high-boned, and wide over the hips, yet the bone not commonly heavy; the hair coarse and long, black or brindled, with white faces. Some are finer in the bone and in the neck, with a good eye and sharp muzzle, and great activity; are hardy, live upon very scanty fare, and fatten with great rapidity when removed to a better soil; they are good milkers. The Kerry cows are excellent in this respect. These last, however, are wild and remarkable leapers. They live, however, upon very little food, and have often been denominated, like those of Pembroke, the poor man's cow."

The other breed is of a larger size. It has much of the blood of the Lancashire or Craven breed, or true Long Horn. Their horns first turn outward, then curve and turn inward. Of each of these kinds, an immense number of both lean and fat stock are annually exported to England.

7. *The Long Horns.*—The Long Horns of England came originally from Craven, in Yorkshire, and derived their name from the length of their horns.

"The improved breed of Leicestershire is said to have been formed by Webster, of Cauley, near Coventry, in Warwickshire. Bakewell, of Dishley, in Leicestershire, afterward got the lead as a breeder, by selecting from the Cauley stock; and the stocks of several other eminent breeders have been traced to the same source.

"The Lancashire breed of long-horned cattle may be distinguished from other cattle by the thickness and firm texture of their hides, the length and closeness of their hair, the large size of their hoofs, and their coarse, leathery, thick necks. They are likewise deeper in their fore-quarters, and lighter in their hind-quarters than most other breeds; narrower in their shape, less in point of weight than the Short Horns, though

better weighers in proportion to their size; and though they give considerably less milk, it is said to yield more cream in proportion to its quantity. They are more varied in color than any other breed; but whatever the color may be, they have in general a white streak along their back, which the breeders term *finched*, and mostly a white spot on the inside of the hough."*

8. *The Short Horn or Durham Breed.*—Durham and York-

Fig. 20.

THE SHORT-HORNED BULL, LORD ERYHOLM.

shire, England, have for ages been celebrated for a breed of short-horned cattle possessing extraordinary value as milkers, "in which quality," the Rev. Henry Barry says, "taken as a breed, they have never been equaled. The cattle so distinguished were always, as now, very different from the improved race. They were generally of large size, thin skinned, sleek haired, bad handlers, rather delicate in constitution, coarse in

---

* Culley.

the offal, and strikingly defective in the substance of girth in the fore-quarters. As milkers they were most excellent, but when put to fatten, as the foregoing description will indicate, were found slow feeders, producing an inferior quality of meat, not marbled or mixed as to fat and lean; the latter sometimes of a very dark hue. Such, too, are the *unimproved* Short Horns of the present day."

The improved Short Horns are even more celebrated as feeders than as milkers, and in other respects differ widely from the original breed.

"The colors of the improved Short Horns," Mr. Youatt says, "are red or white, or a mixture of both;" "no *pure improved Short Horns*," he adds, "are found of any other color but those above named. That the matured Short Horns are an admirable grazier's breed of cattle is undoubted; they are not, however, to be disregarded as milkers; but they are inferior, from their fattening qualities, to many others as workers."

Mr. Dickson, an eminent cattle breeder, thus eloquently describes the Short Horn:

"The external appearance of the short-horned breed is irresistibly attractive. The exquisitely symmetrical form of the body in every position, bedecked with a skin of the richest hues of red, and the richest white approaching to cream, or both colors, so arranged or commixed as to form a beautiful fleck or delicate roan, and possessed of the mellowest touch; supported on clean, small limbs, showing, like those of the race-horse and the greyhound, the union of strength with fineness; and ornamented with a small, lengthy, tapering head, neatly set on a broad, firm, deep neck, and furnished with a small muzzle, wide nostrils, prominent, mildly-beaming eyes, thin, large, biney ears set near the crown of the head and protected in front with semicircularly bent, white, or brownish colored short (hence the name), smooth, pointed horns; all these parts combine to form a symmetrical harmony, which has never been surpassed in beauty and sweetness by any other species of the domesticated ox."

The graziers of Kentucky and other parts of the West have heretofore shown the greatest preference for the Short Horns, but, in their case, they are found to be subject to one serious objection. It is this: while they take on fat so readily when well fed and become so heavy, they are unable to retain it during the long journeys to the Eastern markets, where they generally arrive in too meager a condition to command the price of fat cattle. They require some breed which will be able to carry their fat along with them.*

9. *The Alderney or Jersey Breed.*—This breed of cattle is from Normandy and the Isle of Jersey, and, although small and awkwardly shaped, are much esteemed on account of the richness of their milk, of which, however, the quantity is small. English noblemen keep Alderney cows in their parks to furnish cream for their coffee.

When dried, the Alderney cow fattens with a rapidity that would hardly be thought possible from her gaunt appearance. In color, the Alderney breed is light red, dun, or fawn colored.

10. *The Galloway Breed.*—The Galloway breed of cattle is well known for various valuable qualities, and is easily distinguished by the want of horns. The Galloways are broad across the back, with a very slight curve between the head and the quarters, and broad at the loins, the whole body having a fine round appearance. The head is of moderate size, the ears large and rough, the chest deep, and the legs short. The prevailing color is black. This breed is highly esteemed, as there is no other kind which arrives at maturity so soon; and their flesh is of the finest quality. Their milk is very fine, but is not obtained in very large quantities. It is estimated that 30,000 of these cattle are annually sent out of Galloway.

Another valuable breed of polled (or hornless) cows is bred in Angus, which much resemble, in appearance, those of Galloway; they are, however, rather larger and longer in the leg, flatter sided, and with thinner shoulders.

---

* American Farmer's Encyclopedia.

In Norfolk and Suffolk a hornless breed of cows prevails, which are almost all descended from the Galloways, "whose general form," Mr. Youatt says, "they retain, with some of, but not all, their excellences; they have been enlarged, but not improved, by a better climate and soil. They are commonly of a red or black color, with a peculiar golden circle around the eye. They are taller than the Galloways, but thinner in the chine, flatter in the ribs, and longer in the legs; rather better milkers; of greater weight when fattened, though not fattening so kindly, and the meat is not quite equal in quality."

The Suffolk Dun cow, which is also of Galloway descent, is celebrated as a milker, and there is little doubt is not inferior to any other breed in the quantity of milk which she yields: this is from six to eight gallons per day. The butter produced, however, is not in proportion to the milk. It is calculated that a Suffolk cow produces annually about 1½ cwt. of butter.

The Suffolk Duns derive the last part of their name from their usual pale yellow color. Many, however, are red, or red and white. They are invariably without horns, and small in size, seldom weighing over 700 lbs. when fattened.*

11. *The Cream-Pot Breed.*—This is an American breed, and was originated by Colonel Jaques, of Ten Hills Farm, Somerville, Mass. It is a cross between the Short Horn and the native breed of New England. Mr. Jaques gives the following account of the origin of this famous breed:

"Hearing of cows that produce seventeen pounds of butter each per week, the inquiry arose, why not produce a breed of such cows that may be depended on? This I attempted, and have accomplished. I have made from one of my Cream-Pot cows nine pounds of butter in three days on grass feed only.

"The bull Cœlebs, an imported thorough-bred Durham, and Flora, a heifer of the same breed, and imported, and a native cow, whose pedigree is entirely unknown, comprise the elements of the Cream-Pot breed of cattle. The native cow was

---

* American Farmer's Encyclopedia.

bought in consequence of her superior quality as a milker, giving eighteen quarts a day, and averaging about fifteen. In the month of April the cream of two days' milk produced two and three-fourths pounds of butter, made of two and one-sixteenth quarts of cream, and required but two minutes' churning. Thus much for the mother of the Cream-Pots.

"I have bred my Cream-Pots with red or mahogany colored hair and teats, and gold-dust in the ears, yellow noses and skin, the latter silky and elastic to the touch, being like a fourteen-dollar cloth. My Cream-Pots are full in the body, chops deep in the flank, not quite as straight in the belly, nor as full in the twist, nor quite as thick in the thigh as the Durhams; but in other respects like them. They excel in affording a great quantity of rich cream, capable of being converted into butter in a short time, with little labor, and with a very small proportion of buttermilk, the cream producing more than eighty per cent. of butter. I have changed the cream to butter not unfrequently in one minute, and it has been done in forty seconds."

The late lamented Henry Colman, while Commissioner for the Agricultural Survey of Massachusetts, wrote as follows:

"Mr. Jaques is entitled to great credit for his care and judicious selection in continuing and improving his stock. I have repeatedly seen the cream from his cows, and its yellowness and consistency are remarkable, and in company with several gentlemen of the Legislature, I saw a portion of it converted to butter with a spoon in one minute. The color of Mr. Jaques' stock is a deep red, a favorite color in New England; they are well formed and thrifty on common feed; and if they continue to display the extraordinary properties by which they are now so distinguished, they promise to prove the most valuable race of animals ever known among us for dairy purposes, and equal to any of which we have any information."

III.—POINTS.

Were an ox of fine symmetry and high condition placed before a person not a judge of livestock, his opinion of its

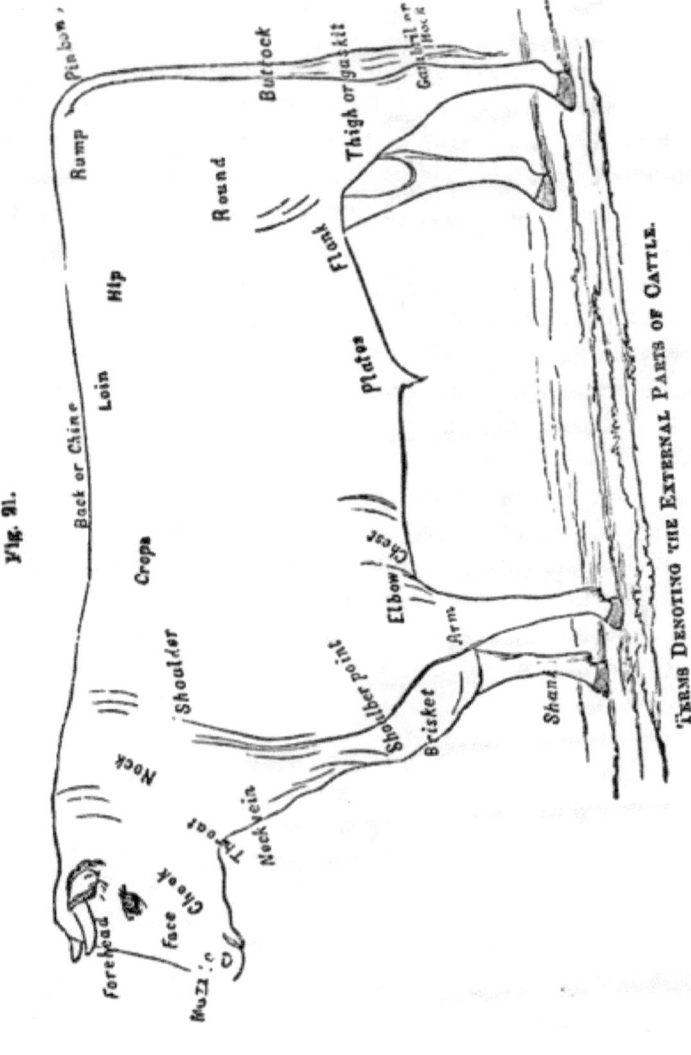

Fig. 91.

TERMS DENOTING THE EXTERNAL PARTS OF CATTLE.

excellences would be derived from a very limited view, and consequently from only a few of its qualities. He could not possibly discover, without tuition, those properties which had chiefly conduced to produce the high condition in which he saw the ox. He would hardly believe that a judge can ascertain merely by the eye, from its general aspect, whether the ox were in good or bad health; from the color of its skin whether it were of a pure or a cross breed; from the expression of its countenance whether it were a quiet feeder; and from the nature of its flesh whether it had arrived at maturity. The discoveries made by the hand of the judge might even stagger belief. He understands the "points" of cattle, and experience enables him to appreciate their individual and aggregate value.

The "points" by which cattle are characterized may profitably be described in detail:

1. The *nose* or *muzzle* in the Durhams or Short Horns should be of a rich cream color. In the Devon, Hereford, and Sussex it is preferred when a clear golden color. A brown or dark color indicates a cross.

2. The *forehead* should be neither narrow nor very broad. The *eye* should be prominent, and the *nostril* between the eye and the muzzle thin, particularly in the Devons.

3. The *horns* should be small, smooth, tapering, and sharp pointed, long or short, according to the breed, and of a white color throughout in some breeds, and tipped with black in others. The shape is less essential than the color.

4. The *neck* should be of medium length, full at the sides, not too deep in the throat, and should come out from the shoulders nearly on a level with the chine.

5. The top of the *plate bones* should not be too wide, but, rising on a level with the chine, should be well thrown back, so that there may be no hollowness behind.

6. The *shoulder point* should lay flat with the ribs, without any projection.

7. The *breast* should be wide and open, projecting forward.

8. The *chine* should lay straight and be well covered with flesh.

9. The *loin* should be flat and wide; almost as wide at the fore as the hinder part.

10. The *hip bones* should be wide apart, coming upon a level with the chine to the setting of the tail.

11. The *tip of the rump* should be tolerably wide, so that the tail may drop to a level between the two points; and the *tail* should come out broad.

12. The *thigh* should not be too full outside nor behind; but the inside or twist should be full.

13. The *back* should be flat and rather thin.

14. The *hind leg* should be flat and thin; the legs of medium length, and the hock rather turning out.

15. The *feet* should not be too broad.

16. The *flank* should be full and heavy when the animal is fat.

17. The *belly* should not drop below the breast, but on a line with it.

18. The *shoulder* should be rather flat, not projecting.

19. The *fore leg* should also be flat and upright, but not fleshy.

20. The *round* should not project, but be flat with the outside of the thigh.

21. The *jaws* should be rather wide.

22. The *ribs* should spring nearly horizontally from the chine and form a circle.

23. The *skin* should be loose, floating, as it were, on a layer of soft fat, and covered with thick, glossy, soft hair.

24. The *expression* of the *eye* and *face* should be calm and complacent.

A writer in the *Farmer's Magazine*, a number of years ago, described what are properly considered the good points of a cow, as exhibited in the Short Horn breed, in the following doggerel lines:

> She's long in her face, she's fine in her horn;
> She'll quickly get fat without cake or corn;
> She's clean in her jaws, and full in her chine;
> She's heavy in flank, and wide in her loin;
> She's broad in her ribs, and long in her rump;
> She's straight in her back, with never a hump;

She's wide in her hip, and calm in her eyes;
She's fine in her shoulders, and thin in her thighs
She's light in her neck, and small in her tail;
She's wide in her breast, and good at the pail;
She's fine in her bone, and silky of skin;
She's a grazier's without, and a butcher's within.

IV.—GENERAL MANAGEMENT.

1. *The Cow-house.*—The cow-house should be a capacious, well-lighted, and well-ventilated building, in which the cows or oxen can be kept dry, clean, and moderately warm. It is a mistaken idea that cattle suffer materially by *dry* cold. It is the wet and the damp walls, yard, and driving rains and fogs of winter, that are so injurious to them. In this respect the Dutch farmers are very particular. They have their cows regularly groomed, and the walks behind them sprinkled with sand.*

As a general thing, our farmers pay too little attention to the health and comfort of their cattle, and especially the cows. In many cases they are kept in a shamefully dirty condition. The floor of their stalls is allowed to be disgustingly filthy, the floors and walls full of vermin, and the hides of the animals covered with dust and dung. It is not only at the expense of their comfort that cattle suffer this neglect, but to the farmer's loss also. When you see a cow rubbing herself against a post, you may depend upon it that the animal is ill kept and requires a good scrubbing. Cattle, as well as horses, are greatly injured by want of proper attention to the cleanliness and ventilation of their habitation. They should stand on a slightly raised platform, which should be well littered with straw, refuse hay, leaves, sawdust, or some other dry material.

For tying up cattle, chains, leather straps, wooden bows, and stanchions are used. The stanchions are the most convenient for the person having charge of the cattle, but, we think, less comfortable for the cattle themselves than the other con-

---

* British Husbandry.

trivances mentioned. A good and cheap stanchion is constructed as follows:

"The sills of the stanchions are of oak joist, six by two inches; the top timbers are of hemlock, of the same dimensions; the stanchions of ash, one and a half by four inches; one of each set of stanchions is pinned between the sills and the corresponding top pieces. From the bottom of the sills to the top of the stanchions is five and a half feet. The slip stanchions are of the same size and material as the first named, but only pinned at the bottom, which allows of their sliding back at the top about sixteen inches, to admit the animal's head; it is then pushed to an upright position and fastened at the top by a drop-button or clapper, which is much more secure than when fastened by pins.

"For oxen and large cows, there is allowed a space for each of three and a half feet; for younger cattle about three feet to each. We have frequently seen the sill and top piece for stanchions made of solid timber, and mortices made for the stanchions. But there is much labor required in morticing, especially the top timber, so as to allow of the sliding back and forward of the slip stanchions. The kind we have attempted to describe can be readily and cheaply made by almost any farmer."*

2. *Feeding.*—While confined to the barn or cow-house and barn-yard, during the cold season, cattle should be fed with the utmost regularity; and a sufficient quantity of nutritious food supplied to keep them in good condition. In this country, hay is the principal common food of our oxen and cows. Roots are too seldom employed in ordinary feeding; and we have no doubt but that the health and, consequently, the condition and value of our cattle would be improved by giving them more turnips, beets, carrots, parsneps, etc., during the winter.

An English writer says: "Supposing a cow to calve early in April or May, there is no keeping to be compared with a sweet pasture for affording the best flavored milk and butter; therefore,

---

* Country Gentleman.

although on a principle of economy I have always recommended the house feeding of a cow (as one acre of good clover will support three cows during the summer, whereas an acre of pasture will but barely suffice for one during the same period, irrespectively of the manure saved by the former management), I make a decided exception where there is no necessity for minutely regarding economy at the expense of the discomfort of the cow, and the inferiority in flavor, if not in quantity, of cream and butter. Yet, even with liberty, and the animal's enjoyment of picking her food as she pleases, there will be necessity in summer for some artificially grown grasses, to supply any deficiency that may occur in the pasture, and provide for the house feeding, when the heat of the sun, the stinging of flies, or the bursting of a storm may render the shade and security of the cow-shed very grateful to your cows. In the early and cold spring, and before the grass has sufficiently sprung up, it is not any kindness to the cow, and it is a decided injury to the ground and vegetation to turn her out; at that season she requires the warmth which her stall affords, and the nourishment that nutritious hay and roots and bran impart."

The following hints from the pen of Henry Colman should be well heeded by every farmer. It is their own fault if American agriculturists do not profit by such truthful warnings.

"The farmers prejudice very greatly their own interest in suffering their milch cows to come out in the spring in low condition. During the time they are dry, they think it enough to give them the coarsest fodder, and that in limited quantities; this, too, at a time of pregnancy, when they require the kindest treatment and the most nourishing food. The calf itself under this treatment of the cow is small and feeble. He finds comparatively insufficient support from his exhausted dam; and the return which the cow makes in milk during the summer is much less than it would be if she came into the spring in good health and flesh. It requires the whole summer to recover what she has lost. The animal constitution can not be trifled with in this way.

"It is so with all livestock, and especially with young animals, at the period of their most rapid growth. They should not be prematurely forced; but, on the other hand, they should not be stinted or checked.

"In the feeding of cattle for market a great deal of practical skill is required, and constant observation of their condition, otherwise they may be surfeited and their appetite destroyed, or their digestive powers be overtasked, and the feed fail of its object.

"The articles usually employed in fattening cattle are hay and Indian meal, or corn and rye meal mixed, or pease and oats, or oats and corn ground together. Besides this, many farmers are in the practice of giving their stall-fed cattle occasionally certain quantities of potatoes. An excellent farmer, of fifty years' experience in the fatting of cattle, is of opinion that potatoes are good feed for fatting cattle in the fall and spring, when the weather is warm; but they do no good in cold weather unless they are cooked. I rely much upon his judgment and experience. The value of potatoes is differently estimated by different individuals; some considering five bushels, others rating four bushels, as equivalent to one bushel of corn."

An extensive cattle-dealer who has tried a variety of mixtures of feed, such as oats, brown-corn seed, etc., prefers Indian meal to every other feed. He disapproves of excessive feeding, and thinks it a great error to give too much. He deems four quarts, with hay, ordinarily enough; and ten quarts a day sufficient for any animal. He feeds twice a day with great regularity. His present cattle have never received over eight quarts per day each; and at first putting up, a much less quantity. He deems it best to reduce their feed of provender a few days before starting for market. He buys his cattle for feeding in the fall; and his present stock averaged in the cost seventy-five dollars per pair.*

---

* American Farmer's Encyclopedia.

"It is sometimes asked," Mr. Colman says, "whether oxen are injured in their growth from being worked. If their strength is prematurely and too severely taxed, or if they are subjected to severe usage, undoubtedly it must prove injurious; but, if otherwise, if reasonably worked and carefully and kindly attended, there is no doubt that their health and growth are promoted by it. It is often matter of inquiry, whether fatting cattle should be kept in close stalls, or be suffered to lie out-doors. The experience of all the farmers whom I have consulted, who have made any trial, is conclusive in this case, in favor of the superior thrift of animals kept constantly in the barn, or turned out only for watering and immediately put up again, over those which are kept in open sheds, or tied up for-feeding only, and at other times allowed to lie in the yard. No exact experiments have been made in this country in relation to this subject; but experiments made abroad lead to the conclusion, that cattle thrive best in a high and equable temperature, so warm as to keep them constantly in a state of active perspiration, and that their thrift is much hindered by an exposure to severe alternations of heat and cold. It is certain, that in order to thrift, cattle can not be made too comfortable; their mangers should be kept clean; their stalls be well littered; and the cattle protected from currents of air blowing through crevices or holes in the floors or the sides of the stables, which prove often much more uncomfortable than an open exposure."

3. *Rearing Calves.*—Many different opinions prevail on the subject of rearing calves. The following plan, detailed by a Western breeder, we deem an excellent one:

"I have my cows so managed that they come in early in spring. I wean the calves after they have drawn the milk two or three times, while I milk at the same time, all clean, that which the calf may not be strong enough to draw. Then I allow the calves nearly all the milk the cows give, for four or six weeks, which gives them a good start; next, I teach them, when two or three weeks old, to eat some little of meal or threshed oats, and lick a little salt; at the same time I let

them have access to some good hay; next, I reduce the quantity of new milk, and give them sweet milk minus the cream, and by degrees teach them to drink coppered milk, feeding ten or twelve together in a trough. This I consider better than milk which is just on a change from sweet to sour. As soon as practicable after there is a good bite of grass, I turn them into pasture, even with the cows, for they know not their dams. I still feed them with milk until about three months old, and all through the season if it can be had. In this wise calves are hearty, learn easy to eat anything which may be offered, and will winter better than calves which have drawn the milk from cows, and have received 'more knocks than nubbins.' They are also more gentle, easier turned to the yoke, or to milk, and are not afraid of their masters; but, on the contrary, learn to know the hands that feed them. By giving them a good chance the first winter, they generally make good thrifty cattle."

4. *Milking.*—In reference to milking, Martin Doyle says: "Cows in general are milked but twice a day, morning and evening; but some of the Durham cows, particularly when in full season and abundantly fed, will require to be milked at noon also. In this case nothing is really gained in the quantity of milk, and its quality is weakened, as twelve hours are required for the due chemical preparation of the milk. Therefore the tendency to this want of retention in a cow is not to be encouraged; the milk should only be drawn off at supernumerary times, if the udder be excessively distended, and the milk flows spontaneously. At each regular time of milking, the contents of the udder should be completely drawn off—the last drop is the richest: when there are two, three, or more cows, the dairy-maid, if she understands her business, will go with a separate vessel and milk the strippings into it until each udder is perfectly dry. This small portion of rich milk will give her more cream than a larger quantity, and she reserves it, if she be a prudent person, for her own tea.

"A cow should be handled with exceeding gentleness, other

wise milking may become an unpleasant or even a painful operation to her. If a cross-grained man or woman, with a vinegar face, handles the teats roughly, and bullies a cow of sensitiveness, she may refuse to let her milk flow, though she would yield to the first touch of a good-tempered person. If the udder be hard, it will require fomentation with lukewarm water and gentle rubbing. It sometimes happens that the teats become sore; in this case an application of sweet oil, after washing the affected part with soap and water, will probably cure it.

"A cow may be milked until within a month of calving, provided the milk does not curdle on being slightly warmed, or possess a salt taste; either would be an indication that no more milk should be taken."

V.—WEIGHT OF LIVE CATTLE.

Experienced drovers and butchers are in the habit, in buying cattle, to estimate their weight on foot. Long experience and much practice enables them to judge with considerable accuracy. They thus have the advantage of the less experienced farmer, who, for this reason, very often comes off "second best" in a bargain. We recommend to them the following rule, by means of which the weight of cattle may be ascertained with a very close approach to the accuracy of the scales.

Rule.—Take a string, put it around the breast, stand square just behind the shoulder-blade, measure on a rule the feet and inches the animal is in circumference; this is called the girth; then, with the string, measure from the bone of the tail which plumbs the line with the hinder part of the buttock; direct the line along the back to the forepart of the shoulder-blade; take the dimensions on the foot rule as before, which is the length; and work the figures in the following manner: Girth of the animal, say 6 feet 4 inches, length 5 feet 3 inches, which multiplied together, makes 31 square superficial feet, and that multiplied by 23, the number of pounds allowed to

each superficial foot of cattle measuring less than 7 and more than 5 feet in girth, makes 713 pounds. When the animal measures less than 9 and more than 7 feet in girth, 31 is the number of pounds to each superficial foot. Again, suppose a pig or any small beast should measure 2 feet in girth and 2 along the back, which multiplied together makes 4 square feet, that multiplied by 11, the number of pounds allowed to each square foot of cattle measuring less than 3 feet in girth, makes 44 pounds. Again, suppose a calf, a sheep, etc., should measure 4 feet 6 inches in girth, and 3 feet 9 inches in length, which multiplied together make 15¼ square feet; that multiplied by 16, the number of pounds allowed to cattle measuring less than 5 feet and more than 3 in girth, makes 265 pounds. The dimensions of girth and length of horned cattle, sheep, calves, and hogs, may be exactly taken in this way, as it is all that is necessary for any computation, or any valuation of stock, and will answer exactly to the four quarters, sinking offal.*

This rule is so simple that any man with a bit of chalk can work it out, and its application will often save the farmer from losses which mere guess work is liable to occasion.

---

* Valley Farmer.

IV.

SHEEP.

Thy flocks the verdant hillside range—Anon.

I.—CHARACTERISTICS.

THE sheep (*Ovis aries*) is naturally a denizen of the hills. Its instincts, even in its domesticated state, attach it to the upland slopes; and when free to do so, it always seeks the highest grounds, where aromatic plants abound and the herbage is less succulent than in the valleys. The wild sheep, like the deer, is found to frequent all those places where saline exudations abound and to lick the salt earth. In its wild state it generally has horns, but these have nearly disappeared in most of the domestic breeds. The female goes with young twenty-one weeks, and usually produces only one at a birth. Twins, however, are not uncommon.

Immense flocks of sheep have been kept by man in all ages, but more generally for their wool and skins than for their flesh; for that is by no means generally relished. The Calmucks and Cossacks still prefer that of the horse and the camel, and the Spaniard, if he can procure other flesh, rarely eats that of the Merino. To a majority of Americans it is an object of dislike, although it is gaining in favor among us. Englishmen consume more mutton than any other people, but the taste for it is of modern origin with them.

The natural age of the sheep, according to Youatt, is about ten years, up to which age they will breed and thrive well; but there are instances of their breeding at the age of fifteen, and living twenty years.

II.—BREEDS.

Specimens of nearly or quite all the valuable breeds of sheep now known may, it is believed, be found in the United States. The principal of these are the Native (so called); the Spanish Merino; the Saxon Merino; the New Leicester or Bakewell; the South-Down; the Cotswold, the Cheviot, and the Lincoln. Between these breeds an almost infinite variety of crosses have taken place; so that, comparatively speaking, few flocks in the United States preserve entire the distinctive characteristics of any one breed, or that can lay claim to purity of blood.*

1. *The Native Breed.*—This name is applied to the common coarse-wooled sheep existing here previous to the importation of the improved breeds. They are, however, of foreign, and mostly of English origin, and probably are the result of the admixture of various breeds. This common stock of sheep, as a distinct family, has nearly disappeared, having been universally crossed, to a greater or less extent, with foreign breeds of later introduction; and especially with the Spanish and Saxon Merinos.

2. *The Spanish Merino Breed.*—Of this excellent breed there have been many importations from France and Spain. There are several varieties of the Merino, differing essentially in form, size, and quality of wool. American Merinos may be classed under three general heads, and are thus described:

"The *first* is a large, short-legged, strong, and exceedingly hardy sheep, carrying a heavy fleece, ranging from medium to fine, somewhat inclined to throatiness, bred to exhibit external concrete gum in some flocks, but not commonly so.

"The *second* general class of American Merinos are smaller than the preceding, less hardy; wool, as a general thing, finer, and covered with a black, pitchy gum on its extremities. The fleece is about one third lighter than in the first class.

"The *third* class, which have been bred mostly at the South, are still smaller and less hardy, and carry lighter and finer

---

* Randall's Sheep Husbandry.

fleeces, destitute of external gum. The sheep and the wool bear a close resemblace to the Saxon, and if not actually mixed with that blood, they have been formed into a similar variety by a similar course of breeding.

"Class *first* are larger and stronger sheep than those originally imported from Spain, and in well-selected flocks or individuals the fleece is of a decidedly better quality."*

The Merino, although a native of a warm climate, becomes readily inured to the greatest extremes of cold, flourishing even so far north as Sweden without degenerating in fleece or form

Fig. 22.

THE SPANISH MERINO.

It is patient, docile, hardy, and long lived. Its flesh, in spite of the prejudice which exists on the subject, is short-grained, and of a good flavor when killed at a proper age. It is longer in coming to maturity than most other breeds, and does not attain its full growth till it is about three years old.†

---

* Randall.   † Transactions of New York State Agricultural Society.

3. *The Saxon Merino Breed.*—The Saxon Merinos are descended from the Spanish, having been imported from Spain into Saxony in 1765. They have been considerably modified by their German breeding, the German shepherds having sacrificed hardiness, and indeed almost everything else, to fineness of staple.

There are very few flocks of pure Saxon sheep in the United States, the importations in several instances having been grade sheep, although sold as pure stock. Most flocks have again been crossed with Native or Spanish Merino sheep or with both; but the mixed breed thus produced, which we may call the American Saxons, have so long been bred toward the Saxons, that their wool equals that of the pure breed. They are hardier than the parent German stock, but still comparatively tender, requiring regular supplies of good food, protection from storms of all kinds, and good shelter in winter. In docility, patience under confinement, late maturity, and longevity, they resemble the Spanish Merinos.*

4. *The New Leicester Breed.*—This celebrated English breed comprehends the most excellent of the breed of Mr. Bakewell, their great improver, and of Mr. Culley's variety or improvement upon it. "The principal recommendations of this breed," Culley says, "are its beauty and its fullness of form; in the same apparent dimensions greater weight than any other sheep; an early maturity and a propensity to fatten equaled by no other breed; a diminution of the proportion of offal, and the return of the most money for the food consumed."

"The wool of the New Leicester," according to Randall, "is long, averaging, after the first shearing, about six inches, and the fleece of the American animal weighs about six pounds. It is of a coarse quality, and is little used in the manufacture of cloths. As a combing wool, however, it stands first, and is used in the manufacture of the finest worsteds, etc."

In England, the mutton of this breed is in great demand, and

---

* Randall.

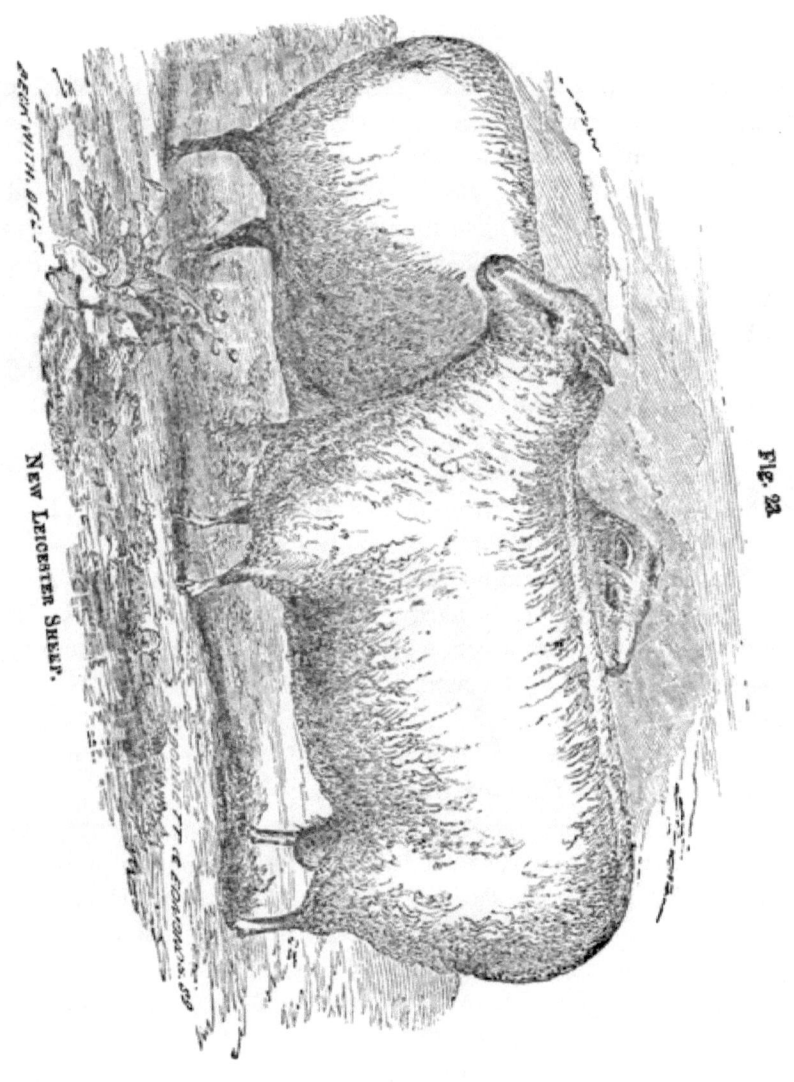

Fig. 24.

New Leicester Sheep.

brings good prices. It is not generally considered a profitable breed in this country, except, perhaps, on rich lowland farms in the vicinity of considerable markets.

5. *The South-Down Breed.* —The South-Down is an upland sheep of medium size, and its wool, in point of length, belongs to the medium class. There has been considerable controversy in reference to the value of the Downs in comparison with the other favorite breeds. Mr. Randall does not rate them very high for wool-bearing. But they are cultivated in England more particularly for their mutton, which in the English markets takes precedence of every other sort.

"The Down is turned off at two years old, and its weight at that age in England is from eighty to a hundred pounds. Notwithstanding its weight, the Down has a patience of occasional short keep, and an endurance of hard stocking equal to any other sheep. It is hardy, healthy, quiet, and docile. It withstands our American winters well. A sheep possessing such qualities must of course be valuable in upland districts in the vicinity of markets."*

Mr. J. C. Taylor, of Holmdel, N. J., in a communication published in the *Country Gentleman*, says:

"I contend that under a high state of management, the South-Downs are a very profitable sheep to keep, in proof of which (for I have the figures) I will cite my now yearling ram. Last July he was worth five dollars to sell for butchering, without anything more than good pasture; he served several ewes from the middle of September to the first of December, which was much against his growth. At seven cents per week, from July to December, say $1 50—cost of keep from December to May 2d, $5 41, making, with his worth in July, a total of $11 91. Had he been a wether I could have sold him on May 2d for $22 for butchering, leaving a clear gain of over $10 at from thirteen to fourteen months old! I ask the stock-raiser and feeder if this is not as profitable as long wools, or

---

* Randall.

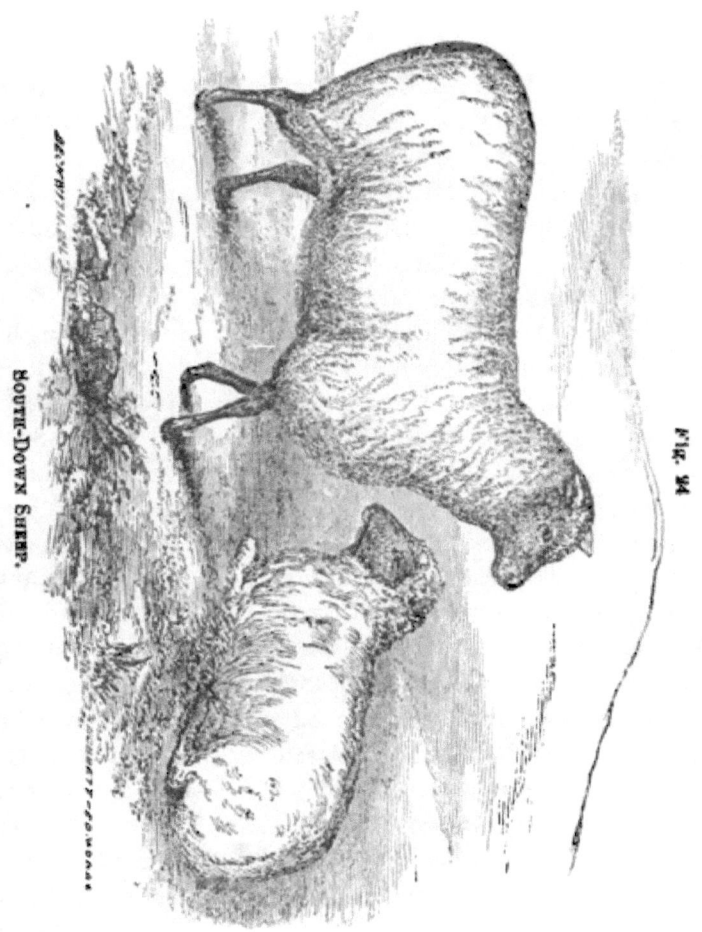

Fig. 24.

SOUTH-DOWN SHEEP.

any other stock? Is it not more profitable? But the Downs are a superior sheep for crossing with common ewes to produce butcher's lambs, superior to *any long wools*.

"A few years ago a Mr. Beers went to Canada and procured a large lot of the Canada Leicester, and many of our farmers were induced by their large size (with their wool on) to buy them. I expected to be driven out of the market with my South-Downs; but at the first county fair (South-Downs having to show against long wool) I made a clean sweep of it, and there has never been one of them shown since. A certain farmer procured one of Mr. Beers' bucks, and also a South-Down; he divided his flock of ewes as nearly as possible between the two bucks; the result was, the half-blood Down lambs were all fat, and sold before any of the half-blood Leicesters were fit for market. This farmer finds the South-Downs *so profitable* that he keeps no other than a South-Down buck."

6. *The Cotswold Breed.*—"The Cotswold," Spooner says, "is a large breed of sheep, with a long and abundant fleece, and the ewes are very prolific and good nurses. They have been extensively crossed with the Leicester sheep, by which their size and fleece have been somewhat diminished, but their carcasses considerably improved, and their maturity rendered earlier. The wool is strong, mellow, and of good color, although rather coarse, from six to eight inches in length, and from seven to eight pounds per fleece. The quality of the mutton is considered superior to the Leicester."

We believe the Cotswolds have not been extensively bred in the United States, although there have been several importations. An improved variety of the Cotswolds, under the name of the New Oxfordshire sheep, have lately attracted considerable attention, and have frequently been successful candidates for prizes offered for the best long-wooled sheep at agricultural shows.

7. *The Cheviot Breed.*—The Cheviot sheep are a peculiar breed, which are kept on the extensive range of the Cheviot Hills. They are described as having "the face and legs gen-

erally white; the eye lively and prominent; the countenance open and pleasing; the ear large, and with a long space from the ear to the eye; the body long; and hence they are called 'long sheep,' in distinction from the black-faced breed. They are full behind the shoulder, have a long, straight back, are round in the rib, and well-proportioned in the quarters; the legs clean and small-boned, and the pelt thin, but thickly covered with fine, short wool; they possess very considerable fattening qualities, and can endure much hardship, both from starvation and cold."*

We have no acquaintance with this breed. There are probably but few of them in this country. Mr. Randall speaks very disparagingly of those which had fallen under his observation, but which may have not been fair specimens of their breed.

8. *The Lincoln Breed.*—Culley described the old breed of Lincolnshire sheep, half a century ago, as having "no horns, white faces, long, thin, and weak carcasses; the ewes weighing from 14 to 20 lbs. per quarter, the three-year old wethers from 20 to 30 lbs.; thick, rough, white leg, large bones, thick pelts, and long wool, from 10 to 18 inches, and weighing from 8 to 14 lbs. per fleece, and covering a slow-feeding, coarse-grained carcass of mutton." Culley, however, ran into the opposite extreme; if the Lincolnshire farmers bred only for the wool, he regarded only the mutton. A cross between the two produced a very profitable and much improved animal.

III.—CHOICE OF BREED.

"In selecting a breed for any given locality," Mr. Randall says, "we are to take into consideration, first, the feed and climate, or the surrounding natural circumstances; and second, the market facilities and demands. We should then make choice of that breed which, with the advantages posssessed, and under all the circumstances, will yield the greatest net value of marketable product.

---

* American Farmer's Encyclopedia.

"Rich lowland herbage, in a climate which allows it to remain green during a large portion of the year, is favorable to the production of large carcasses. If convenient to markets where mutton finds a ready sale at good prices, then all the conditions are realized which call for a mutton as contradistinguished from a wool-producing sheep. Under such circumstances, the choice should undoubtedly, in my judgment, rest between the improved English varieties—the South-Down, the New Leicester, and the improved Cotswold or New Oxfordshire. In deciding between these, minor and more specific circumstances are to be taken into account."

For wool-growing purposes he thinks the Merino "possesses a marked and decided superiority over the best breeds and families of coarse-wooled sheep;" and its inferiority as a mutton sheep, he thinks is not so great as is generally supposed.

IV.—GENERAL MANAGEMENT.

The following hints are all condensed from Randall's excellent work on Sheep Husbandry, to which the reader who may desire further details is referred.

1. *Barns, Sheds, etc.*—"Humanity and economy both dictate that sheep be provided with shelters to lie under nights, and to which they can resort *at will*. In our severe winter storms it is sometimes necessary, or at least by far the best, to feed under shelter for a day or two. It is not an uncommon circumstance, in New York and New England, for snow to fall to the depth of twenty or thirty inches, within twenty-four or forty-eight hours, and then to be succeeded by a strong and intensely cold west or northwest wind of two or three days' continuance,* which lifts the snow, blocking up the roads, and piling huge drifts to the leeward of fences, barns, etc. A flock without shelter will huddle closely together, turning their backs to the storm, constantly stepping and thus treading down the snow as it rises about them. Strong, close-coated sheep do not

---

* These terrible wind-storms are of much longer continuance in many parts of New England.

seem to suffer as much from the cold, for a period, as would be expected; but it is next to impossible to feed them enough or half enough, under such circumstances, without an immense waste of hay—entirely impossible, without racks. The hay is whirled away in an instant by the wind; and even if racks are used, the sheep leaving their huddle where they were kept warm and even moist by the melting of the snow in their wool, soon get chilled and are disposed to return to their huddle. Imperfectly filled with food, the supply of animal heat is lowered, and at the end of the second or third day the feeble ones have sunk down hopelessly, the yearlings and oldish ones have received a shock which nothing but careful nursing will recover them from, and even the strongest have suffered an injurious loss in condition.

"The simplest and cheapest kind of shed is formed by poles or rails, the upper ends resting on a strong horizontal pole supported by crotched posts set in the ground. It may be rendered rain-proof by pea-haulm, straw, or pine boughs.

"In a region where lumber is very cheap, planks or boards (of sufficient thickness not to spring downward, and thus open the roof), battened with slabs, may take the place of the poles and boughs; and they would make a tighter and more durable roof. If the lower ends of the boards or poles are raised a couple of feet from the ground, by placing a log under them the shed will shelter more sheep.

"These movable sheds may be connected with hay-barns, 'hay-barracks,' stacks, or they may surround an inclosed space with a stack in the middle. In the latter case, however, the yard should be *square*, instead of *round*, on account of the divergence in the lower ends of the boards or poles, which the round form would render necessary."

2. *Feeding-Racks.*—"When the ground is frozen, and especially when covered with snow, the sheep eats hay better on the ground than anywhere else. When the land is soft, muddy, or foul with manure, they will scarcely touch hay placed on it. It should then be fed in racks.

"These are of various forms. Fig. 25 gives the common box rack in the most general use in the North. It is ten feet long, two and a half wide, the lower boards a foot wide, the upper ones about ten inches, the two about nine inches apart,

Fig. 25.

BOX RACK.

and the corner posts three by three, or three and a half by two and a half inches. The boards are spiked on these posts by large flat-headed nails wrought for the purpose, and the lower edges of the upper boards and the upper edges of the lower ones are rounded so they shall not wear the wool off from the sheep's necks. The lower boards and the opening for the heads should be two or three inches narrower for lambs. If made of light wood, as they should be, a man standing in the inside and middle of one of these racks, can easily carry it about—an important desideratum. Unless over-fed, sheep waste very little hay in them."

An improvement upon the common box rack has holes eight inches wide, nine inches high, and about eighteen inches apart, instead of the continuous opening represented in the foregoing cut; but it is a little more expensive.

3. *Feeding.*—"In Germany great stress is laid on variety in the winter fodder, and elaborate systems of feeding are given. Variations of dry fodder are well enough, but hundreds and thousands of Northern flocks receive nothing but ordinary hay, consisting mainly of timothy (*Phleum pratense*), some red and white clover (*Trifolium pratense et repens*), and frequently a sprinkling of June or spear-grass (*Poa pratensis*), during the entire winter. Others receive an occasional fodder of cornstalks and straw, and some farmers give a daily feed of grain

through the winter. Where hay is the principal feed, it may be well, where it is convenient, to give corn-stalks (or 'blades') every fifth or sixth feed, or even once a day; or the daily feed, *not of hay*, might alternate between blades, pea-straw, straw of the cereal grains, etc. Should any other fodder besides hay be the principal one, as, for example, corn-blades or peahaulm, each of the *other* fodders might be alternated in the same way. It is mainly, in my judgment, a question of convenience with the flock-master, provided a *proper supply of palatable nutriment within a proper compass* is given. Hay, clover, properly cured pea-haulm, and corn-blades are palatable to the sheep, and each contain the necessary supply of nutriment in the quantity which the sheep can readily take into its stomach. Consequently, from either of these, the sheep can derive its entire subsistence. Sheep should not run or be fed in yards with any other stock.

"The expediency of feeding grain to store sheep in the winter depends upon circumstances. Remote from markets, it is generally fed by the holders of large flocks. Oats are commonly preferred, and they are fed at the rate of a gill a head per day. Some feed half the same amount of (yellow) corn. Fewer sheep—particularly lambs, yearlings, and crones—get thin and perish, where they receive a daily feed of grain; they consume less hay, and their fleeces *are increased in weight*. On the whole, therefore, it is considered good economy. Where no grain is fed, three daily feeds of hay are given. It is a common and very good practice to feed greenish cut oats *in the bundle*, at noon, and give but two feeds of hay—one at morning and one at night. A few feed greenish cut peas in the same way. In warm, thawing weather, when sheep get to the ground, and refuse dry hay, a little grain assists materially in keeping up their strength and condition. This may furnish a useful hint for many parts of the South. When the feed is shortest in winter, in the South, there are many localities where sheep would get enough grass to take off their appetite for dry hay, but not *quite* enough to keep them in prime con-

dition. A moderate daily feed of oats or peas, placed in the depository racks, would keep them strong, in good plight for the lambing season, and increase their weight of wool.

"Ruta-bagas, Irish potatoes, etc., make a good substitute for grain, as an extra feed for grown sheep. I prefer the ruta-baga to the potato in equivalents of nutriment. I do not consider either of them, or any other root, as good for lambs and yearlings as an equivalent in grain. Sheep may be *taught* to eat nearly all the cultivated roots; this is done by withholding salt from them, and then feeding the chopped root a few times rubbed with just sufficient salt to induce them to eat the root to obtain it, but not enough to satisfy their appetite for salt before they have acquired a taste for the roots.

"If there is one rule which may be considered more imperative than any other in sheep husbandry, it is that the utmost regularity be preserved in feeding. First, there should be regularity as to the *times* of feeding. However abundantly provided for, when a flock are foddered sometimes at one hour and sometimes at another—sometimes three times a day and sometimes twice—some days grain and some days none—*they can not be made to thrive*. They will do far better on *inferior keep*, if fed with strict regularity. In a climate where they require hay three times a day, the best times for feeding are about sunrise in the morning, at noon, and an hour *before dark* at night. Unlike cattle and horses, sheep do not eat well *in the dark*, and therefore they should have time to consume their food before night sets in. Noon is the common time for feeding grain or roots, and is the best time if but two fodderings of hay be given. If the sheep receive hay three times, it is not a matter of much consequence with which feeding the grain is given, only that the practice be uniform.

"It is also highly essential that there be regularity preserved in the *amount* fed. The consumption of hay will, it is true, depend much upon the weather. The keener the cold, the more sheep will eat. In the South, much would also depend upon the amount of grass obtained. In many places a light,

daily foddering would suffice—in others, a light foddering placed in the depository racks once in two days would answer the purpose. In the steady cold weather of the North, the shepherd readily learns to determine about how much hay will be consumed before the next foddering time; and this is the amount which should, as near as may be, be *regularly* fed. In feeding grain or roots there is no difficulty in preserving *entire regularity*, and it is vastly more important than in feeding hay. Of the latter a sheep will not over-eat and surfeit itself; of the former it will. And if not fed grain to the point of surfeiting, but still over-plenteously, it will expect a like amount at the next feeding, and failing to receive it will pine for it and manifest uneasiness. The effect of such irregularity on the stomach and system of *any* animal is bad, and the sheep suffers more from it than any other animal. I would much rather that my flock receive no grain at all than that they should receive it without regard to regularity in the amount. The shepherd should be required to *measure* out the grain to sheep in all instances—instead of *guessing* it out—and to measure it to each separate flock.

"In the North the grass often gets very short by the 10th or 15th of November, and it has lost much of its nutritiousness from repeated freezing and thawing. At this time, though no snow has yet fallen, it is best to give the sheep a light, daily foddering of bright hay, or a few oats in the bundle. Given thus for the ten or twelve days which precede the covering of the ground by snow, fodder pays for itself as well as at any other time during the year."

4. *Salt.*—"Salt, in my judgment, is indispensable to the health of sheep, particularly in the summer; and I know not a flock-master among the hundreds, nay, thousands with whom I am acquainted, who differs with me in this opinion. It is common to give it once a week while the sheep are at grass.

"It is still better to give them free access to salt at all times by keeping it in a covered box, open on one side."

5. *Water.*—"Water is not indispensable in the summer pastures, the dews and the succulence of the feed answering as a

substitute. But my impression is decided that free access to water is advantageous to sheep, particularly to those having lambs; and I should consider it a matter of importance, on a sheep farm, to arrange the pastures, if practicable, so as to bring water into each of them."

6. *Shade.*—" No one who has observed with what eagerness sheep seek shade in hot weather, and how they pant and apparently suffer when a hot sun is pouring down on their nearly naked bodies, will doubt that, both as a matter of humanity and utility, they should be provided, during the hot summer months, with a better shelter than that afforded by a common rail fence. Forest trees are the most natural and best shades, and it is as contrary to utility as it is to good taste to strip them entirely from the sheep-walks. A strip of stone wall or close board fence on the south and west sides of the pasture will form a passable substitute for trees; but in the absence of all these, and of buildings of any kind, a shade can be cheaply constructed of poles and brush, in the same manner as the sheds of the same materials for winter shelter already described."

7. *Lambs.*—" Lambs are usually dropped in the North from the first to the fifteenth of May. In the South, they might safely come earlier. It is not expedient to have them dropped when the weather is cold and boisterous, as they require too much care; but the sooner the better after the weather has become mild, and the herbage has started sufficiently to give the ewes that green food which is required to produce a plentiful secretion of milk. It is customary in the North to have fields of clover, or the earliest of grasses, reserved for the early spring feed of the breeding ewes; and if these can be contiguous to their shelters, it is a great convenience—for the ewes should be confined in the latter, on cold and stormy nights, during the lambing season.

"If warm and pleasant, and the nights are warmish, I prefer to have the lambing take place in the pastures. I think sheep are more disposed to own and take kindly to their lambs thus,

than in the confusion of a small inclosure. Unless particularly docile, sheep in a small inclosure crowd from one side to another when any one enters, running over young lambs, and pressing them severely, etc. Ewes get separated from their lambs, and then run violently round from one to another, jostling and knocking them about. Young and timid ewes get separated from their lambs, and frequently will neglect them for an hour or more before they will again approach them. If the weather is severely cold, the lamb, if it has never sucked, stands a chance to perish. Lambs, too, when just dropped, in a *dirty* inclosure, in their first efforts to rise, tumble about, and the membrane which adheres to them becomes smeared with dirt and dung—and the ewe refuses to lick them dry, which much increases the hazard of freezing.

"Lambs should be weaned at four months old. It is better for them and much better for their dams. The lambs when taken away should be put for several days in a field distant from the ewes, that they may not hear each other's bleatings. The lambs when in hearing of their dams continue restless much longer, and they make constant and frequently successful efforts to crawl through the fences which separate them. One or two tame old ewes are turned into the field with them, to teach them to come at the call, find salt when thrown to them, and eat grain, etc., out of troughs when winter approaches.

"The lambs when weaned should be put on the freshest and tenderest feed. I have usually reserved for mine the grass and clover sown, the preceding spring, on the grain fields which were seeded down.

"The dams, on the contrary, should be put for a fortnight on short, dry feed, to stop the flow of milk. They should be looked to once or twice, and should the bags of any be found much distended, the milk should be drawn and the bag washed for a little time in cold water. But on short feed they rarely give much trouble in this particular. When properly dried off they should be put on good feed to recruit, and get in condition for winter."

8. *Emasculation and Docking.*—"These should usually precede washing, as at that period the oldest lambs will be about a month old, and it is safer to perform the operations when they are a couple of weeks younger. Dry, pleasant weather should be selected. Castration is a simple and safe process. Let a man hold the lamb with its back pressed firmly against his breast and stomach, and all four legs gathered in front in his hands. Cut off the bottom of the pouch, free the testicle from the inclosing membrane, and then draw it steadily out, or clip the cord with a knife, if it does not snap off at a proper distance from the testicle. Some shepherds draw both testicles at once with their *teeth*. It is common to drop a little salt into the pouch. Where the weather is very warm, some touch the end of the pouch (and that of the tail, after that is cut off) with an ointment, consisting of tar, lard, and turpentine. In ninety-nine cases out of a hundred, however, they will do just as well, here, without any application.

"The tail should be cut off, say one and a half inches from the body, with a chisel on the head of a block, the skin being slid up toward the body with a finger and thumb, so that it will afterward cover the end of the stump. Severed with a knife, the end of the tail being grasped with one of the hands in the ordinary way, a naked stump is left which takes some time to heal.

"It may occur to some unused to keeping sheep, that it is unnecessary to cut off the tail. If left on, it is apt to collect filth, and if the sheep purges, it becomes an intolerable nuisance.

9. *Washing.*—"This is usually done here about the first of June. The climate of the Southern States would admit of its being done earlier. The rule should be to wait until the water has acquired sufficient warmth for bathing, and until cold rains and storms, and cold nights are no longer to be expected.

10. *Shearing.*—"It is difficult, if not impossible, to give intelligible practical instructions which would guide an entire novice in skillfully shearing a sheep. Practice is requisite. The

following directions from the American Shepherd* are correct, and are as plain, perhaps, as they can be made:

"'The shearer may place the sheep on that part of the floor assigned to him, resting on its rump, and himself in a posture with one (his right) knee on a cushion, and the back of the animal resting against his left thigh. He grasps the shears about half-way from the point to the bow, resting his thumb along the blade, which affords him better command of the points. He may then commence cutting the wool at the brisket, and proceeding downward, all upon the sides of the belly to the extremity of the ribs, the external sides of both thighs to the edges of the flanks; then back to the brisket, and thence upward, shearing the wool from the breast, front, and both sides of the neck—but not yet the back of it—and also the poll or fore-part, and top of the head. Now the "jacket is opened" of the sheep, and its position and that of the shearer is changed, by being turned flat upon its side, one knee of the shearer resting on the cushion, and the other gently pressing the fore-quarter of the animal, to prevent any struggling. He then resumes cutting upon the flank and rump, and thence onward to the head. Thus one side is complete. The sheep is then turned on to the other side, in doing which great care is requisite to prevent the fleece from being torn, and the shearer acts as upon the other, which finishes. He must then take his sheep near to the door through which it is to pass out, and neatly trim the legs, and leave not a solitary lock anywhere as a harbor for ticks. It is absolutely necessary for him to remove from his stand to trim, otherwise the useless stuff from the legs becomes intermingled with the fleece wool. In the use of the shears, let the blades be laid as flat to the skin as possible, not lower the points too much, nor cut more than from one to two inches at a clip, frequently not so much, depending on the part and compactness of the wool.'

"Cold storms sometimes destroy sheep, in this latitude, soon

---

* Pages 179, 180.

after shearing—particularly the delicate Saxons. I have known forty or fifty perish out of a single flock, from one night's exposure. The remedy, or rather the preventive, is to house them, or in default of the necessary fixtures to effect this, to drive them into dense forests. I presume, however, this would be a calamity of rare occurrence in the 'sunny South.'"*

V.—VALUE OF SHEEP TO THE FARMER.

The following suggestive remarks are from the *Country Gentleman*, and are worthy of every reader's attention:

"Sheep are profitable to the farmer, not only from the product of wool and mutton, but from the tendency which their keeping has to improve and enrich his land for all agricultural purposes. They do this:

"1. By the consumption of food refused by other animals in summer; turning waste vegetation to use, and giving rough and bushy pastures a smoother appearance, and in time eradicating wild plants so that good grasses and white clover may take their place. In this respect sheep are of especial value to pastures on soils too steep or stony for the plow. In winter, the coarser parts of the hay, refused by horses and cows, are readily eaten by sheep, while other stock will generally eat most of that left by these animals.

"For these reasons, among others, no grazing farm should be without at least a small flock of sheep, for it has been found that as large a number of cattle and horses can be kept with as without them, and without any injury to the farm for other purposes. A small flock, we said—perhaps half a dozen to each horse and cow would be the proper proportion. A va-

---

* Sheep Husbandry; with an Account of the Different Breeds and General Directions in regard to Summer and Winter Management, Breeding, and Treatment of Diseases. With Portraits and other Engravings. By Henry S. Randall. New York: A. O. Moore. This work is bound with "Youatt on the Sheep," under the general title of "The Shepherd's Own Book," and the volume should be in the hands of every one who would make sheep-breeding his principal business.

riety of circumstances would influence this point; such as the character of the pasturage, and the proportion of the same fitted and desirable for tillage.

"2. Sheep enrich land by the manufacture of considerable quantities of excellent manure. A farmer of long experience in sheep husbandry, thought there was no manure so fertilizing as that of sheep, and (of which there is no doubt) that none dropped by the animal upon the land suffered so little by waste from exposure. A German agricultural writer has calculated that the droppings from one thousand sheep during a single night would manure an acre sufficiently for any crop. By using a portable fence, and moving the same from time to time, a farmer might manure a distant field with sheep at less expense than that of carting and spreading barn manure.

"The value of sheep to the farmer is much enhanced by due attention to their wants. Large flocks kept together are seldom profitable, while small assorted flocks always pay well, if fed as they should be. To get good fleeces of wool, and large, healthy lambs from poor neglected sheep, is impossible. It is also true that the expense of keeping is often least with the flocks that are always kept in good condition. The eye and thought of the owner are far more necessary than large and irregular supplies of fodder. Division of the flock and shelter, with straw and a little grain, will bring them through to spring pastures in far better order than if kept together, with double rations of hay, one half of which is wasted by the stronger animals, while the weak of the flock pick up but a scanty living, and oftentimes fail to get that through the whole winter.

"We commend this subject to the consideration of our correspondents; it is one which needs greater attention on the part of the farming public."

VI.—AFFECTION OF THE EWE.

The Ettrick Shepherd tells the following story of the continued affection of the ewe for her dead lamb:

"One of the two years while I remained on the farm at Wil-

lenslee a severe blast of snow came on by night, about the latter end of April, which destroyed several scores of our lambs, and as we had not enow of twins and odd lambs for the mothers that had lost theirs, of course we selected the best ewes and put lambs to them. As we were making the distribution, I requested of my master to spare me a lamb for a ewe which he knew, and which was standing over a dead lamb in the end of the hope, about four miles from the house. He would not let me do it, but bid me let her stand over her lamb for a day or two, and perhaps a twin would be forthcoming. I did so, and faithfully she did stand to her charge. I visited her every morning and evening for the first eight days, and never found her above two or three yards from the lamb; and often as I went my rounds, she eyed me long ere I came near her, and kept stamping with her foot, and whistling through her nose, to frighten away the dog. He got a regular chase twice a day as I passed by; but however excited and fierce a ewe may be, she never offers any resistance to mankind, being perfectly and meekly passive to them.

"The weather grew fine and warm, and the dead lamb soon decayed; but still this affectionate and desolate creature kept hanging over the poor remains with an attachment that seemed to be nourished by hopelessness. It often drew tears from my eyes to see her hanging with such fondness over a few bones, mixed with a small portion of wool. For the first fortnight she never quitted the spot; and for another week she visited it every morning and evening, uttering a few kindly and heart-piercing bleats; till at length every remnant of her offspring vanished, mixing with the soil, or wafted away by the winds."

V.

SWINE.

*Where oft the swine, from ambush warm and dry,
Bolt out and scamper headlong to their sty.—Bloomfield.*

I.—NATURAL HISTORY.

THE hog (*Suidæ sus* of Linnæus), according to Cuvier, belongs to "the class *Mammalia*, order *Pachydermata*, genus *Suidæ* or *sus*."

Professor Low remarks, that "the hog is subject to remarkable changes of form and characters, according to the situation in which he is placed. When these characters assume a certain degree of permanence, a breed or variety is formed; and there is no one of the domestic animals which more easily receives the characters we desire to impress upon it. This arises from its rapid powers of increase, and the constancy with which the characters of the parents are reproduced in the progeny.

There is no kind of livestock that can be so easily improved by the breeder and so quickly rendered suited to the purposes required; and the same characters of external form indicate in the hog a disposition to arrive at early maturity of muscle and fat as in the ox and the sheep. The body is long in proportion to the limbs, or, in other words, the limbs are short in proportion to the body; the extremities are free from coarseness; the chest is broad and the trunk round. Possessing these characteristics, the hog never fails to arrive at early maturity, and with a smaller consumption of food than when he possesses a different conformation."

The wild boar, which was undoubtedly the progenitor of all the European varieties, and also of the Chinese breed, was for-

merly a native of the British Islands, and very common in the forests until the time of the civil wars in England.

The wild hog is now spread over the temperate and warmer parts of the old continent and its adjacent islands. His color varies with age and climate, but is generally a dusky brown with black spots and streaks. His skin is covered with coarse hairs or bristles, intersected with soft wool, and with coarser and longer bristles upon the neck and spine, which he erects when in anger. He is a very bold and powerful creature, and becomes more fierce and indocile with age. From the form of his teeth he is chiefly herbivorous in his habits, and delights in roots, which his acute sense of smell and touch enables him to discover beneath the surface. He also feeds upon animal substances, such as worms and larvæ which he grubs up from the ground, the eggs of birds, small reptiles, the young of animals, and occasionally carrion; he even attacks venomous snakes with impunity.

The female produces a litter but once a year, and in much smaller numbers than when domesticated. She usually carries her young for four months or sixteen weeks.

In a wild state the hog has been known to live more than thirty years; but when domesticated he is usually slaughtered for bacon before he is two years old, and boars killed for brawn seldom reach to the age of five. When the wild hog is tamed, it undergoes the following among other changes in its conformation. The ears become less movable, not being required to collect distant sounds. The formidable tusks of the male diminish, not being necessary for self-defense. The muscles of the neck become less developed, from not being so much exercised as in the natural state. The head becomes more inclined, the back and loins are lengthened, the body rendered more capacious, the limbs shorter and less muscular; and anatomy proves that the stomach and intestinal canals have also become proportionately extended along with the form of the body. The habits and instincts of the animal change; it becomes diurnal in its habits, not choosing the night for its search of food;

is more insatiate in its appetite, and the tendency to obesity increases.

The male forsaking its solitary habits, becomes gregarious, and the female produces her young more frequently, and in larger numbers. With its diminished strength and power of active motion, the animal also loses its desire for liberty. These changes of form, appetites, and habits, being communicated to its progeny, a new race of animals is produced, better suited to their altered condition. The wild hog, after it has been domesticated, does not appear to revert to its former state and habits; at least the swine of South America, carried thither by the Spaniards, which have escaped to the woods, retain their gregarious habits, and have not become wild boars.*

II.—OPINIONS RESPECTING THE HOG.

From the various allusions to the hog in the writings of the ancient Greeks and Romans, it is plain that its flesh was held in high esteem among those nations. The Romans even made the breeding, rearing, and fattening pigs a study, which they designated as *Porculatio*.

Varro states that the Gauls produced the largest and finest swine's flesh that was brought into Italy; and according to Strabo, in the reign of Augustus, they supplied Rome and nearly all Italy with gammons, hog-puddings, and sausages. This nation and the Spaniards appear to have kept immense droves of swine, but scarcely any other kind of livestock. . . . . In fact, the hog was held in very high esteem by all the early nations of Europe; and some of the ancients have even paid it divine honors.†

On the other hand, swine's flesh has been held in utter abhorrence by the Jews since the time of Moses, in whose laws they were forbidden to make use of it as food. The Egyptians also and the followers of Mohammed have religiously abstained from it. Paxton, in his "Illustrations of Scripture," says:

---

* American Farmer's Encyclopedia. † Youatt.

"The hog was justly classed by the Jews among the vilest animals in the scale of animated nature; and it can not be doubted that his keeper shared in the contempt and abhorrence which he had excited. The prodigal son in the parable had spent his all in riotous living, and was ready to perish through want, before he submitted to the humiliating employment of feeding swine."

"Swine," Heroditus says, "are accounted such impure beasts by the Egyptians, that if a man touches one even by accident, he presently hastens to the river and, in all his clothes, plunges into the water. For this reason swine-herds alone of the Egyptians are not allowed to enter any of their temples; neither will any one give his daughter in marriage to one of that profession, nor take a wife born of such parents, so that they are necessitated to intermarry among themselves."

The Brahminical tribes of India share with the Jews, Mohammedans, and Egyptians this aversion to the hog. The modern Copts, descendants of the ancient Egyptians, rear no swine, and the Jews of the present day abstain from their flesh as of old.

It was Cuvier's opinion that "in hot climates the flesh of swine is not good;" and Mr. Sonnini remarks that "in Egypt, Syria, and even the southern parts of Greece, this meat, though very white and delicate, is so far from being firm, and is so overcharged with fat, that it disagrees with the strongest stomachs. . It is therefore considered unwholesome, and this will account for its proscription by the legislators and priests of the East. Such abstinence was doubtless indispensable to health under the burning suns of Arabia and Egypt." How is it under the burning suns of Carolina and Georgia?

III.—BREEDS.

The various breeds which have been reared by crosses between those procured from different countries are so numerous, that to give anything like a detailed description of them would fill a large volume. We shall refer to only a few of the more important of them.

1. *The Land Pike.*—The old common breed of the country, sometimes called "land-pikes," may be described as "large, rough, long-nosed, big-boned, thin-backed, slab-sided, long-leg-

Fig. 26.

The Land Pike Hog.

ged, ravenous, ugly animals." Speaking of this race, A. B. Allen says: "No reasonable fence can stop them, but, ever restive and uneasy, they rove about seeking for plunder; swilling grunting, rooting, pawing; always in mischief and always destroying. The more a man possesses of such stock the worse he is off." But this breed is rapidly disappearing. Crosses between the land-pike and the Chinese or the Berkshire producing a fine animal, the original breed is being very generally improved.

2. *The Chinese Breed.*—This breed was introduced into this country from China some forty years ago. The Chinese hog is small in limb, round in body, short in head, and very broad in cheek. When fattened, it looks quite out of proportion, the head appearing to be buried in the neck, so that only the tip of the nose is visible. It has an exceedingly thin skin and fine bristles.

The pure-blooded Chinese hog has been bred to only a limited extent in the United States, on account of the smallness of its size (it seldom attaining more than two hundred and fifty pounds), and its lack of hardiness in a cold climate. In this last respect, however, it is well adapted to the South. Crossed

with the native hog it forms an excellent breed, which we may call the improved China breed. Hogs of this mixed breed are various in color—black, white, spotted, and gray and white; they are longer in body than the pure Chinese breed; small in the head and legs; broad in the back; round in the body; the hams well let down; skin thin; flesh delicate and finely flavored. They are easy keepers; small consumers; quiet in disposition; not disposed to roam; and when in condition may be kept so upon grass only.

3. *The Berkshire Breed.*—This was one of the earliest improved of the English breeds, and is deemed by many the most excellent of all the varieties at present known. It is certainly the most widely distributed and most generally approved. It is a breed which is distinguished by being, in general, of a tawny white, or rufous-brown color, spotted with black or brown; head well placed, large ears, generally standing forward, though sometimes hanging over the eyes; body thick, close, and well made; legs short, small in the bone; coat rough and curly, wearing the appearance of indicating both skin and flesh of a coarse quality. Such, however, is not the case, for they

Fig. 27.

THE BERKSHIRE HOG.

have a disposition to fatten quickly: nothing can be finer than the bacon, and the animals attain to a very great size.

The Berkshires, from which most of the present American stock has sprung, were imported in 1822. The breed has spread very rapidly over the country.

Fig. 28.

THE SUFFOLK HOG.

4. *The Suffolk Breed.*—The improved Suffolk breed originated in a cross between the original Suffolk hog and the Chinese. It is a very valuable breed, but much smaller in size than the Berkshire. The Suffolks are thick through the shoulders, very handsomely proportioned in body, and possessing beautiful hams. Their color is either white or light flesh color, when of the pure breed, and they are indeed an ornament to the farm.

It is said that they are less inclined to cutaneous diseases than numerous others, and do not, under any circumstances, produce that strong, musky flavored pork we sometimes find in market. They are not a gross, unwieldy animal, generally ranging from two hundred and fifty to three hundred pounds

weight at twelve months of age, which latter weight they seldom exceed. They are clean feeders, and require much less than any other breed known.

For large hogs, a cross between these and the Berkshire is very desirable, and is preferred by Western breeders; but for a small breeder, or for family use, the pure Suffolks are preferable.*

5. *The Essex Breed.*—The Essex hogs are mostly black and white, the head and hinder parts being black and the back and belly white. The most esteemed Essex breeds, Youatt says, are entirely black, and are distinguished by having small teat-like appendages of the skin depending from the under part of the neck. They have smaller heads than the Berkshire hogs, and long, thin, upright ears; short bristles; a fine skin; good hind quarters, and a deep, round carcass. They are also small boned, and their flesh is delicately flavored. They produce large litters, but are reputed bad nurses.

Fig. 29.

THE ESSEX HOG.

6. *The Chester Breed.*—This breed originated in Chester County, Pennsylvania, and is not so widely known as, according

---

* Country Gentleman.

to all accounts, it deserves to be. A correspondent of the *Country Gentleman* gives the following account of the Chester hog:

"The Chester hog is the result of continued careful breeding and judicious crossing in this county during the last thirty-five or forty years. The first impulse to this improvement, it is said, was the importation of a pair of handsome hogs from China, some forty years since, by a sea-captain then residing in this vicinity. Of late years, however, many of our breeders have been laboring to bring the Chester hog up to an acknowledged standard of excellence—to define its points, and make it as distinctive in character, and as easily recognized, as a Berkshire or Suffolk. Their efforts, we think, have been successful.

"The genuine Chester is a pure white, long body and square built, with small, fine bone, and will produce a greater weight of pork, for the amount of food consumed, than any other breed yet tried among us. A very important characteristic of the breed is, that it will *readily fatten at any age*. Many hogs, it is well known, will not fatten while they are growing, or until they have reached their full size.

"The average weight of the Chester stock, at sixteen months old, is from 500 to 600 lbs., and when kept till two years old, they frequently run up to 700 and 800 lbs. Our spring pigs, when killed the following fall, weigh from 300 to 400 lbs., which is considered the most desirable weight for pork—producing hams of a more salable size and better quality. As a general rule, our farmers do not care to have their hogs weigh over 350 to 400 lbs. To reach this weight at nine months old, our hogs, of course, must be well fed. The Chester is not different from other stock in this respect—to thrive well, it must be well taken care of.

"Experiments have been made in crossing the Chester with other breeds—such as the Berkshire, Suffolk, etc., and the result has been an inferior stock to the pure Chester. It *does* improve the Berkshires to cross them with the Chester, but we have found no advantage in crossing the Chester with any other."

IV.—POINTS.

"There is evidently much diversity in swine in different circumstances and situations. Like other descriptions of stock, they should be selected with especial reference to the nature of the climate, the keep, and the circumstances of the management under which the farm is conducted. The chief points to be consulted in judging of the breeds of this animal are the form or shape of the ear, and the quality of the hair. The pendulous or lop ear, and coarse, harsh hair, are commonly asserted to indicate largeness of size and thickness of skin; while erect or prick ears show the size to be smaller, but the animals to be more quick in feeding.

"In the selection of swine, the best formed are considered to be those which are not too long, but full in the head and cheek; thick and rather short in the neck; fine in the bone; thick, plump, and compact in the carcass; full in the quarters, fine and thin in the hide; and of a good size according to the breed, with, above all, a kindly disposition to fatten well and expeditiously at an early age. Depth of carcass, lateral extension, breadth of the loin and breast, proportionate length, moderate shortness of the legs, and substance of the gammons and fore-arms, are therefore absolute essentials. These are qualities to produce a favorable balance in the account of keep, and a mass of weight which will pull the scale down. In proportion, too, as the animal is capacious in the loin and breast, will be generally the vigor of his constitution; his legs will be thence properly distended, and he will have a bold and firm footing on the ground."*

V.—FEEDING.

Have regular hours for feeding your hogs; nothing is more important. Irregularity irritates the digestive organs, and prevents the system from receiving the full benefit of the meal when it does come. Do not give them too much food at once,

---

* American Farmer's Encyclopedia.

as they are apt to gorge themselves; or, if any be left in the trough, to return to it frequently till it is all gone. In both cases their digestive organs, and consequently their ability to fatten, are impaired.

Swine will eat animal food, but it is not favorable to the flavor of their flesh, and should always be withheld while they are fattening.

Pigs always eat more when first put up to fatten than they do afterward, therefore the most nutritious food should be reserved till they are getting pretty fat.

In reference to fattening the hog, a writer in the *Boston Cultivator* remarks:

"If circumstances are favorable, he is inclined to lay up such a supply of fat during autumn as would render it unnecessary for him to undergo much exercise or exposure during inclement weather. With plenty of *lard oil* to keep his lamp burning, he would prefer dozing in a bed of leaves in the forest while the ground is covered with snow, rather than to *grub* daily for a living. He fattens most rapidly in such a state of the atmosphere as is most congenial to his comfort—neither too hot nor too cold; hence the months of September and October are best for making pork. The more agreeable the weather, the less is the amount of food required to supply the waste of life.

"Against fattening hogs so early in the season, it may be objected that Indian corn, the crop chiefly depended on for the purpose, is not matured. Taking everything into consideration, it may be better to begin to feed corn before it is ripe, or even at the stage of considerable greenness. After the plant has blossomed it possesses a considerable degree of sweetness; hogs will chew it, swallow the juice, and leave nothing but the dry fibrous matter, which they eject from their mouths when no more sweetness can be extracted. They thrive on this fodder, and will continue to eat it till the nutriment is concentrated in the ear, and then they will eat the cob and grain together till the cob gets hard and dry. Farmers who have practiced this mode of feeding consider it more advantageous

than to leave the whole crop to ripen, unless they have a supply of old corn to feed with. Even in the latter case, it is questionable whether hogs will not do better on corn somewhat green than they would on hard corn, unground. True, it is not necessary that corn should be fed unground, but much is fed in this condition, no doubt at a loss.

"In many parts of the country, swine are fed considerably on articles which are not readily marketable, as imperfect fruits, vegetables, etc. Where such articles are used, cooking them is generally economical. A mixture of squashes (either summer or winter squashes), pumpkins—the nearer ripe the better—potatoes, beets, and apples, boiled or steamed, and a fourth or an eighth of their bulk of meal stirred in while the mass is hot, forms a dish on which hogs will fatten fast. If skimmed milk or whey can be had, the cooked food may be put with it into a suitable tub or vat, and a slight degree of fermentation allowed to take place before the whole is fed out. The animals will eat it with avidity, and probably derive more benefit from it than if it had not been fermented. Articles which are of a perishable nature should be used first in fattening swine, in order to prevent waste and turn all the products of the farm to the best account.

"Another quite important advantage of early feeding is the less trouble in regard to cooking the food and keeping it in proper condition to feed out. The cooking may be done out of doors, if convenience of feeding would be promoted by it, and there is no expense or trouble to guard the food against freezing."

The manner of fattening hogs, where Indian corn is used, as at the South and West, is to put them up in large, open pens on the ground, without litter and without shelter. Here they are left to burrow and sleep in mud and mire, exposed to all weathers, consuming, probably, before they get "ripe fat," one third if not half more than would be necessary were they sheltered in a warm pen, with clean litter, clean water, and rich food in abundance, free alike from exposure and excitement.

An ample supply of good drinking water should be kept within the reach of every animal.

VI.—THE PIGGERY.

In constructing a piggery, reference should be had to the comfort of the animals as well as to convenience in feeding them. It should be large, airy, and well-ventilated, and should have (at least in a large establishment) conveniences for cooking their food. It should by all means be comfortable and clean. It has been generally believed that the hog is naturally a filthy animal, delighting in mud and mire. This is certainly, in part at least, untrue. No animal more fully appreciates a clean, dry bed. To illustrate the value of cleanliness, a gentleman in Norfolk (England) put up six pigs of almost exactly the same weight, and all in equal health to fatten; treated them all, except in one particular, exactly alike, giving equal quantities of the same food to each for seven weeks. Three of these pigs were left to shift for themselves, so far as cleanliness is concerned, while the other three were carefully curried, brushed, and washed. The latter consumed, during the seven weeks, less food by five bushels than the former, and yet, when killed, weighed more by thirty-two pounds on an average. [For a plan for a piggery, see "The House."]

VI.

IMPROVEMENT OF BREEDS.

Like produces like.

I.—SELECTION.

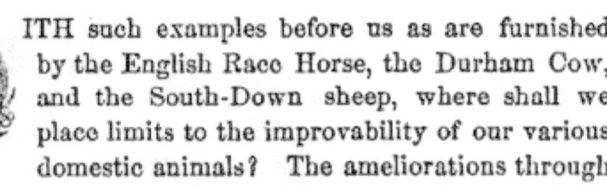

WITH such examples before us as are furnished by the English Race Horse, the Durham Cow, and the South-Down sheep, where shall we place limits to the improvability of our various domestic animals? The ameliorations through which these improved breeds have been established were not accidental. They took place according to the fixed laws of animal life, brought to bear by the intelligence of man upon special points and for special objects. Other breeds even better than these may be produced by similar means. Bakewell, Culley, Seabright, Jaques, Knight, and other distinguished breeders and improvers of stock, have made use of no patented or secret process. What they have done, any intelligent farmer may do by the use of the same easily available means. To furnish a few hints in reference to these means is the purpose of this chapter.

In setting about originating a new breed of any particular species of animal, the first grand point is the selection of sire and dam. This must be made with reference to the particular qualities to which you desire to give prominence, as well as to the general excellence of constitution, form, and disposition which should distinguish the species. Thus Colonel Jaques, in originating the Cream-Pot breed of cows, already referred to, had the dairy and not the butcher in view, and took his measures accordingly. The results of a continued selection of

breeders with reference to their qualities as milkers has been the establishment of a permanent breed distinguished probably above all others as dairy cows. So the sheep breeders of England, having the production of mutton as their principal object, have produced the New Leicester, the South-Down, and the New Oxfordshire breeds, distinguished for form, size, flavor, and fattening qualities; while the Spanish and German breeders of Merinos, caring only for the wool, have given their breeds pre-eminently excellent fleeces. Breeding carefully for a few generations with a distinct purpose in view, will not fail to produce astonishing and satisfactory results.

"The alteration," Sir John Seabright says, "which may be made in any breed of animals by selection can hardly be conceived by those that have not paid some attention to the subject."

To breed in the most successful manner, the male and female should be taken when they are in the highest state of health, and when all the powers and attributes which are wished for and which it is designed to propagate are in the most complete order and state of perfection.

II.—IN-AND-IN BREEDING.

It is a well-established fact in human physiology that the intermarriage of near relatives tends to both physical and mental degeneracy. Analogy would lead us to infer that the same results must follow close breeding among the lower animals; and facts, we think, prove conclusively that this is the case. Youatt, high authority on this subject, says:

"Breeding in-and-in has many advantages to a certain extent. It may be pursued until the excellent form and quality of the breed are developed and established. It was the source whence sprung the fine cattle and sheep of Bakewell, and the superior cattle of Colling; but disadvantages attend breeding 'in-and-in,' and to it must be traced the speedy degeneracy, the absolute disappearance, of the new Leicester cattle, and in the hands of many an agriculturist, the impairment of consti-

tution and decreased value of the new Leicester sheep and the short-horned beasts. It has therefore become a kind of principle with the agriculturist to effect some change in his stock every second or third year; and that change is most conveniently effected by introducing a new bull or ram. These should be as nearly as possible of the same sort, coming from a similar pasturage and climate, but possessing no relationship, or at most a very distant one, to the stock to which he is introduced." These remarks apply to all descriptions of live-stock. In cattle, as well as in the human species, defects of organization and permanent derangements of function obtain, and are handed down when the relationship is close.

III.—CROSSING.

It is by judicious crossing of breeds that some of our best varieties of domestic animals have been obtained. A cross between a superior and an inferior breed results in a progeny superior to the latter, and, for a particular use, climate, or locality, often better than the former. Thus the cross between the English thorough-bred horse and the inferior mare of the common breed of New England gave us the Morgan breed, which for all the common purposes for which a horse is used is superior to the thorough-bred animal himself.

In breeding from stock with qualifications of different descriptions and in different degrees, the breeder will decide what are indispensable or desirable qualities, and will cross with animals with a view to establish them. His proceeding will be of the "give-and-take" kind. He will, if necessary, submit to the introduction of a trifling defect in order that he may profit by a great excellence; and between excellences perhaps somewhat incompatible he will decide which is the greatest, and give it the preference.

The following account of the way in which the new French breed of sheep, La Chamois, was originated, throws light upon an important principle in breeding; namely, that the influence of the male upon the offspring will be the stronger the purer

and more ancient in the first place his own race may be; and in the next place, the less resistance is offered by the female through the possession of those qualities of purity and long descent which are so valuable in the sire.

The French writer says: "With a view to the experiment proposed, it was necessary to procure English rams of the purest and most ancient race, and unite with them French ewes of the modern breeds, or rather of mixed blood forming no distinct breed at all. It is easier than one might have supposed to combine these conditions. On the one hand, I selected some of the finest rams of the New-Kent breed, regenerated by Goord. On the other hand, we find in France many border countries lying between distinct breeds, in which districts it is easy to find flocks participating in the two neighboring races. Thus, on the borders of Berry and La Sologne one meets with flocks originally sprung from a mixture of the two distinct races that are established in those two provinces. Among these, then, I chose such animals as seemed least defective, approaching, in fact, the nearest to, or rather departing the least from, the form which I wished ultimately to produce. These I united with animals of another mixed breed, picking out the best I could find on the borders of La Beauce and Touraine, which blended the Tourangelle and native Merino blood of those other two districts. From this mixture was obtained an offspring combining the four races of Berry, Sologne, Touraine, and Merino, without decided character, without fixity, with little intrinsic merit certainly, but possessing the advantage of being used to our climate and management, and bringing to bear on the new breed to be formed, an influence almost annihilated by the multiplicity of its component elements.

"Now what happens when such mixed-blood ewes are put to a pure New-Kent ram? A lamb is obtained containing fifty hundredths of the purest and most ancient English blood, with twelve and a half hundredths of four different French races, which are individually lost in the preponderance of

English blood, and disappear almost entirely, leaving the improving type in the ascendant. The influence, in fact, of this type was so decided and so predominant, that all the lambs produced strikingly resembled each other, and even Englishmen took them for animals of their own country. But what was still more decisive, when these young ewes and rams were put together they produced lambs closely resembling themselves, without any marked return to the features of the old French races from which the grandmother ewes were derived. Some slight traces only might perhaps be detected here and there by an experienced eye. Even these, however, soon disappeared, such animals as showed them being carefully weeded out of the breeding flock. This may certainly be called '*fixing a breed*,' when it becomes every year more capable of reproducing itself with uniform and marked features."

IV.—ADDITIONAL HINTS.

Farmers, like men in other branches of business, have an eye on the profits of their industry; and the more intelligent of them are now fully convinced of the fact, that with proper care and protection the improved and finer breeds do give a greater product with the same amount of food than the inferior and coarser breeds. It costs but little if any more to keep a cow that will give a large quantity of rich milk than one that does not pay for her food; strong, active horses are far more profitable than poor, lazy ones; a bushel of corn will make twice as much pork when fed to a Berkshire or a Suffolk as to a Land-Pike or Racer, and the best sheep will yield double the wool and bring triple the price of the poorer kinds.

Now every farmer may, in a few years, make great improvement in his stock by selecting his best animals to breed from, with an occasional infusion of fresh blood from other flocks and herds (without reference to any of the celebrated improved breeds), combined with proper attention to their feeding and general management; but unless he has a particular taste for breeding animals, and unusual facilities for the business, he will

find it more convenient and cheaper to make an infusion of the improved blood into his stock, choosing such as is best fitted for his purpose. A bull or a ram of one of the best breeds will soon, if judiciously managed, make a great change for the better in his stock.

Another important fact must be borne in mind. "Improved breeds owe their present degree of perfection, whatever it may be, only to the skill which has been exercised in their selection, breeding, and management for a number of generations and a long series of years. This attention, we learn from the extract above, *must be continued* if we wish to retain the valuable qualities that it has placed within our reach; and careful attention to the selection, the wants, the comfort, and the health of one's stock is thus shown to be not only the dictate of economy for the time being, but a matter of importance in the future, from the influence it exerts on the progeny as well as on the parent. Improvements may be *bred out* as fast or faster than it can be *bred in*. Until the average of care which our farm stock now receives becomes much greater, it may be inexpedient to advise the maintenance of a herd or flock of pure improved blood for ordinary farm purposes; but, by beginning with grades—employing the services of an improved male to engraft upon "native" stock—and by degrees acquiring the habit of paying closer attention to their necessities and comforts, not only will our cattle and sheep be gradually and fundamentally bettered, but the farmer will be preparing to avail himself of breeds already rendered capable of giving, with proper attention, the greatest product for a specified amount of food; and animals bred to this point will then come into his hands to be improved, not to be deteriorated."*

---

* Country Gentleman.

VII.

DISEASES AND THEIR CURE.

Throw physic to the dogs.—Shakspeare.

I.—HYGIENE.

"THROW physic to the dogs," if you will, but, be assured, they are quadrupeds of too much good sense to swallow it; and the other domestic animals will hardly take, except under compulsion, what their canine companions and protectors thus reject. You will find less difficulty in forcing it down the throats of their more frequently diseased and oftener doctored masters.

A large portion of almost every work on domestic animals is taken up with directions for the treatment of their diseases. Our limits do not permit us to dwell long on this point, nor do we deem it necessary.

In their wild state, animals are ordinarily subject to few if any diseases. They live according to the laws of their being— live naturally and healthfully, and, unless they meet a violent death at the hands of man or of some of their natural enemies, die a natural death. Our domestic animals, as they are generally managed, live under conditions less favorable to health, and sometimes, although with comparative infrequency, get sick. The fault is generally in the keeper or breeder, and not in the animal or in the conditions inseparable from its domestic state. With animals, as with men, disease arises from some infringement of the organic laws; but their masters, and not themselves, are responsible for the infringement. When they get sick, however, in consequence of the false conditions under

which they are forced to live, man adds insult to injury by forcing his nauseous and poisonous drugs down their reluctant throats. If they recover in spite of both the disease and the remedy, drugs get the credit.

Well, let those use drugs who have faith in them, either in the treatment of themselves, their families, or their domestic animals; but the reader who looks in this little manual for directions for their use will be disappointed. We can not conscientiously give them.

Animals born of well developed and perfectly healthy parents (and none but perfectly healthy and well developed animals should ever be permitted to become breeders) may almost universally *be kept in perfect health.* With a sufficient quantity of wholesome food, pure water, protection against storms and cold in winter, complete ventilation and perfect cleanliness in their habitations, and general attention to their comfort and health, there will be little call for medical treatment of any kind; and in the rare cases which may occur, we would trust mainly to Nature, co-operating with her as we could by means of diet, air, exercise, and water, on the same principles precisely that are applied in the treatment of human beings without drugs.

The Water-Cure or Hydropathic system has not yet been extensively applied to animals; but so far as it has been adopted, it has produced the most satisfactory results; and for the benefit of such of our readers as may have lost their faith in drugs, and desire to make a trial of a more rational method, we lay before them the following essay, kindly furnished for this work by that distinguished physician and writer, R. T. Trall, M.D., Principal of the New York Hygeio-Therapeutic College.

II.—WATER-CURE FOR DOMESTIC ANIMALS.
BY R. T. TRALL, M.D.

The habits of domestic animals being, on the whole, less unphysiological than those of human beings, their diseases are, as a necessary consequence, less numerous and less complicated. They may all be grouped under the head of *fevers, inflam-*

mations, spasms or colics, fluxes, eruptions, and glandular affections. And for all of these disorders we are satisfied that proper attention to hygiene, as understood by the term Hydropathy or Water-Treatment, is as much superior to drug medication as it has proved to be in the case of human beings similarly affected.

Fever is easily known by the languor and lassitude which the animal manifests, with great indisposition to exercise, followed by chills or shivering, and this succeeded by preternatural heat on the surface, loss of appetite, furred tongue, frequent or hard or bounding pulse, etc. The animal should be placed in a clean, quiet, well-ventilated room, protected from currents of cold air in winter or the scorching rays of the sun in summer, and the temperature should be kept at a uniform and moderate degree continually.

When the skin becomes very hot, it should be washed or bathed all over, and a blanket or two immediately applied, so as to promote moderate perspiration. Or the wet sheet may be applied, taking care to cover it well with blankets, so as to arrest chilliness. When the sheet becomes quite warm, it should be removed, and the surface washed with cold water; and if the fever heat continues, it may be re-applied for an hour at a time, two or three times a day, until the morbid heat is entirely subdued.

The same general plan of treatment, with a slight modification, applies to all inflammatory complaints. With domestic animals as with human beings, the organs most liable to acute inflammation are the lungs and the bowels, and the only specialty of treatment in these affections, in addition to the general plan applicable to the constitutional disturbance we call fever, is the continual application of wet cloths well covered with dry ones to the chest or bowels, as either is the seat of the inflammation, and the employment of copious enemas of tepid water to free the bowels.

Spasmodic diseases of all kinds, and all the varieties of colic, are the results of local obstruction caused by over-exertion,

over-heating, or something improper or indigestible in the food. Grain, and especially Indian meal, fed to a horse while in a state of great heat or great fatigue from violent exertion, is frequently the immediate cause of colic and spasms. In these cases the animal should have his abdomen fomented with wet cloths applied as warm as can be borne; warm water should be given the animal to drink, or poured down his throat from a bottle, and copious enemas of warm water should be administered.

Fluxes—as *diarrhea, dysentery, cholera, influenza, catarrh*, etc.—are the indications of a general obstruction of the system or impurity of the fluids, with an effort at depuration in a particular direction. The usual practice of checking the discharge suddenly by pungents, stimulants, and astringents is always injurious and generally dangerous. On the contrary, the action of the surface should be restored by bathing, with friction or the dripping-sheet, and all irritating matters removed from the stomach and bowels by means of warm and tepid water, as in the case of colics. There will be no danger from the discharges if the cause is removed, and if it is not removed, the sudden suppression of the evacuations may terminate in a worse inflammation or speedy death.

Affections of the skin and glands are only to be cured by purifying the whole mass of blood. To repel an eruption from the surface, or rather a glandular tumor, is not curing the animal; indeed, it is only changing an external disease to an internal one. Thus attention to a pure diet, to fresh air, and to clean apartments, each and all are essential to recovery. Many of these *cachexies*, as they are called in medical books, originate from the effluvia of their own excretions, as in cases where the urine and feces are permitted to accumulate in the stalls, or under the floors of the stables.

VIII.

POULTRY.

Also fowls were prepared for me.—Nehemiah v. 18.

I.—THE DOMESTIC FOWL.

NOBODY knows when or by whom fowls were first domesticated. There are at most only two or three allusions to them in the Old Testament, and these are of doubtful import. In our motto, for instance, the word fowls may mean simply birds.

In the time of Aristotle, who wrote three hundred and fifty years before Christ, however, they were evidently common; for he speaks of them as familiarly as a naturalist of the present day. Everybody is familiar with the beautiful allusions to them in the New Testament.

The wild origin of our domestic fowl is entirely unknown. The race, like that of the Dodo, is probably extinct. The Wild Turkey will sooner or later share the same fate.

Crested or top-knotted fowls appear to have been unknown to the ancients. The earliest notice of them occurs in Aldrovandi, who speaks of a hen with "a crest like a lark."

Domestic fowls now abound in all warm and temperate climates, but disappear as we approach the poles. They were found in abundance on the islands of the Pacific Ocean by their earliest discoverers. How they got there nobody knows. Probably in the same way that their human inhabitants found their insular homes. Ellis, in his "Polynesian Researches," says: "The traditions of the people state that fowls have existed on the islands (Tahiti) as long as the people; that they

came with the first colonists by whom the islands were peopled; or that they were made by Taarva at the same time that men were made."

The courage of the cock is emblematic, his gallantry admirable, his sense of discipline and subordination most exemplary. See how a good game-cock of two or three years' experience will, in five minutes, restore order into an uproarious poultry-yard! He does not use harsh means of coercion when mild will suit the purpose. A look, a gesture, a deep chuckling growl, gives the hint that turbulence is no longer to be permitted; and if these are not effectual, severer punishment is fearlessly administered. . . . . His politeness to females is as marked as were Lord Chesterfield's attentions to old ladies, and much more unaffected. Nor does he merely act the agreeable dangler; when occasion requires, he is also their brave defender, if he be good for anything.

"The hen is deservedly the acknowledged pattern of maternal love. When her passion of philoprogenitiveness is disappointed by the failure or subtraction of her brood, she will either go on sitting till her natural powers fail, or will violently kidnap the young of some other fowl and insist upon adopting them."*

The varieties of the domestic fowl are almost numberless, but only a few of them are worthy of more than a mere mention here. Among these we give the first place to—

1. *The Spanish Fowl.*—The thorough-bred Spanish fowl is entirely black, so far as feathers are concerned, with a greenish metallic luster. The combs of both the male and the female are very large and of a brilliant scarlet; that of the hen droops over on one side. Their most singular feature is a large white patch, or ear-lobe, on the cheek, which in some specimens extends over a large part of the face. It is a fleshy substance, similar to the wattle, and is small in the hens but large and conspicuous in the cocks, giving them a very striking appear-

---

* Rev. Edmund Saul Dixon.

ance. There are few, if any, handsomer fowls than the genuine Spanish; although some that are called by that name, but are really nameless mongrels, are ugly enough for scarecrows.

The hens are great layers, being in this respect, we believe, superior to every other breed. Their eggs are very large, quite white, and of a peculiar shape, being quite thick at both ends, although tapering off a little at each. A correspondent of the *Country Gentleman*, relating his experience with them, says: "My last year's June pullets commenced to lay in December, and the first of February all of my Spanish hens laid more or less. I got, in the six months, from the first of March to the first of September, eighty-five dozen of eggs from seven pullets, and I now get from four to six eggs per day; and my honest

Fig. 30.

THE SPANISH FOWL.

conviction is, that the true Black Spanish hen will lay from 'five to ten per cent.' more weight of eggs than any other breed."

On the other hand, it must be confessed that these Spanish dames are not good mothers or nurses, even when they do sit, "which," as Dixon remarks, "they will not often condescend to do." This last trait of character will prove a recommendation rather than otherwise with those who care for eggs rather than chickens. When the latter are wanted, it is better to place the eggs under a hen of another and more motherly breed—a Dorking, for instance.

The Spanish fowls bear confinement very well; are not large eaters; grow rapidly; mature early; and are only excelled for the table by the Game fowl and the Dorking. The average weight of the mature birds is about six pounds for the male and five for the female.

It is important, but somewhat difficult at present, to procure the true, unmixed, white-faced Black Spanish breed.

There is another breed called the Gray or Speckled Spanish, but, however excellent they may be (and they are highly spoken of), they are probably a mixed breed.

2. *The Dorking Fowl.*—The Dorking takes its name from a town in Surrey County, England, where it is supposed to have originated.

The Dorkings are divided into the Colored and the White varieties; the former including the Gray, Speckled, Spangled, Japanned, etc. These are not *permanent* varieties, however, as they can not be bred true to color. The Gray and Spangled comprise the more common forms in which the Colored Dorking family is presented to us.

The White Dorking is a smaller-framed bird than the Gray, and should be perfectly white in plumage, bill, and legs. They should have rose-combs. They are less hardy than the colored variety, and not well adapted to a northern climate.

The Dorking is a fowl of rare beauty, large in size, symmetrical in form, and often gorgeous in plumage. Its flesh is white, firm, and of excellent flavor; and for the general purposes of the table it is inferior to none, although, as regards flavor alone, the Game fowl would perhaps take precedence.

As layers, the Dorking hens take high rank, but are, we think, inferior to the Spanish. They are persistent sitters, and make excellent mothers and nurses. The editor of the *American*

Fig. 31.

THE DORKING FOWL.

Agriculturist says: "A little knowledge in keeping them [the Dorkings] justified us in pronouncing them entitled to the same rank among barn-yard fowls that the Short Horns have taken among cattle; and years of experience in breeding them have confirmed us in this opinion."

John Giles, a well known poultry breeder of Woodstock, Conn., expresses the following opinion: "After forty-odd years' experience with the gallinaceous tribes, I say that, in my humble opinion, no breed of fowls will compare with the true Dorking as good mothers, sitters, and layers, giving eggs in abundance, chickens easily reared, and which come to perfection sooner than any other poultry. The flesh is of a delicate white, fine in the grain, and delicious flavor. The Black Span-

ish is only second to the true Dorking, in not raising their own young, seldom or ever wanting to sit; but what they lose in that point is more than made up by the abundance of eggs By some they are called the everlasting layers; eggs large; flesh and skin beautifully white and juicy; chickens grow rap idly."

A cross between the Dorking and the Game fowl is greatly esteemed, and is thought to be more profitable than the thorough-bred Dorking.

The possession of the fifth claw is generally considered as an essential characteristic of the Dorking, but it is not always present, and might and should be "bred out." The weight of the Dorking at maturity varies from five to eight pounds.

3. *The Polish Fowl.*—The origin of this family of fowls is entirely unknown. They do not exist in Poland at the present time, and there is no evidence that they were ever known there; but this is a matter of small moment. Their beauty

Fig. 82.

THE POLISH FOWL.

and excellence are undisputed. The large top-knot is one of the principal characteristics of the Polish fowl, and is conspicuous in all its varieties.

The varieties of the Polish or Poland fowl are numerous; but the principal ones are the White-Crested Black, the Golden Spangled, and the Silver Spangled.

In the White-Crested Black Poland cock the plumage, with the exception of the crest, should be uniformly black, with rich metallic tints of green. The shorter crest feathers at the base of the bill are black, the rest of the purest white. The beak and legs are generally black. The same colors are required in the hen. Their form and bearing are remarkably good. The cock should weigh from five to five and a half pounds, and the hen about four pounds.

The Golden Spangled and the Silver Spangled Polands are splendid birds. "The beautiful regularity of their markings, the vivid contrasts in their colors, together with their unique appearance generally, entitle them to the first rank among the more ornamental varieties."

The Polands, and especially the Black variety, are generally but not invariably great layers, commencing early in the spring, and seldom wanting to sit till late in the summer, if at all. They can not always be depended upon to hatch a *clutch* of chickens, even when they manifest a desire to sit, frequently deserting the nest after five or six days' occupation.*
They are not quite so hardy as some other breeds, but with a fair degree of attention are easily reared. As a table fowl, the Polish is among the best.

4. *The Hamburg Fowl.*—Of the Hamburgs there are several varieties. The Silver Penciled, known also as the Bolton Gray, have the plumage white, with the exception of the wings and tail, which are furred with black. The average weight of the cock is about four and a half pounds. The hen usually weighs about a pound less. The Golden Penciled Hamburg

---

* Wingfield.

differs from the Silver Penciled chiefly in the ground color of its plumage, which is a yellowish buff or yellowish bay, and in being rather larger. The legs of both these varieties should be blue. The Silver Spangled and Golden Spangled differ from the Penciled sorts, in having black, circular, oval, or crescent-shaped spangles on the tail and wing, instead of bars. They are somewhat larger than the Penciled birds and have darker

THE SILVER SPANGLED HAMBURG FOWL.

legs. The Black Hamburg has a plumage of a uniformly rich, glossy-green black.

All the Hamburgs are beautiful fowls, rich in plumage and fine in form; great layers (the eggs, however, are small); seldom desire to sit; and are good for the table, falling but little below the best varieties in this respect, although not so large as some others.

They are impatient of confinement, and to do well must have

126   DOMESTIC ANIMALS.

Fig. 34.

THE GOLDEN SPANGLED HAMBURG FOWL.

a wide range of grassy lawn or pasture. Of the different varieties we prefer the Golden Spangled, but others may choose differently.

5. *The Dominique Fowl.*—This is a very common breed in

Fig. 35.

THE DOMINIQUE COCK.

this country, but none the less valuable or beautiful on that account.

"The prevailing and true color of the Dominique fowl is a lightish ground, barred crosswise, and softly shaded with a slaty-blue, as indicated in the portrait of the cock figured on the previous page. The comb is variable, some being single, while others are double—most, however, are single. The iris, bright orange; feet, legs, and bill, bright yellow; and some light flesh color. We prefer the yellow legs and bill, and consider them well worthy of promotion in the poultry-yard.

"We seldom see bad hens of this variety; and take them 'all-in-all,' we do not hesitate in pronouncing them *one* of the *best* and most profitable fowls, being hardy, good layers, careful nurses, and affording excellent eggs, and the quality of their flesh highly esteemed. The hens are not large, but plump and full breasted. The eggs average about two ounces each, and are of porcelain whiteness."*

6. *The Leghorn Fowl.*—The Leghorns are believed to be cousins of the Spanish, whom they resemble in general form. They have been considerably experimented with in this country, and are highly extolled by some breeders; but the general verdict is that they are inferior to the Spanish.†

7. *The Shanghai Fowl.*—The Shanghai fowl was originally brought from the northern part of China, particularly about the city of Shanghai, from which it takes its name. It is the common domestic fowl of that part of the country.

The Shanghai cock is a large, bold, upright bird, strongly distinguished for the length, loudness, hoarseness, and awkwardness of his half guttural crow. Most of the sub-varieties

---

* Country Gentleman.

† A correspondent of one of the agricultural papers, however, gives the following testimony in their favor: "I have kept in different inclosures six of the most approved varieties of fowls, for four months (from the 1st of April to the present)—have registered the number of eggs laid by each variety every day, and the Leghorns have laid almost three eggs to any other bird's one, not excepting the far famed Black Spanish."—R. W. PEARSALL, *Harlem, N Y*

have large, single, serrated combs, the top running considerably beyond its point of attachment to the head. His neck is about nine inches long, and is somewhat arched; wings short, rounded outward, their shoulders concealed in the breast-feathers, and their tips covered by the body-feathers and the saddle-hackle. His breast is broad, but wanting in fullness; the thighs are wide apart, large, comparatively short, smooth in

Fig. 36.

THE SHANGHAI FOWL.

some, in others heavily feathered quite down to the knees; shanks should be short, and, with the booted, more or less feathered down the outer edge, quite to the end of the outer toe; the stern is densely covered with long downy feathers, technically called "fluff," well rounded out; the hackle, both of neck and saddle, is long and abundant; while the tail is short and sometimes covered by the long saddle-feathers. The

weight of a full-grown bird is from ten to twelve pounds, while a few have weighed more. The hen agrees in general character with that of her liege lord, but is two or three pounds lighter.

The legs of both sexes should be yellow, though we have seen some very fine white birds with a greenish-blue leg, and superior black ones with dark legs.

The principal sub-varieties of the Shanghai family are the White, the Buff, the Cinnamon, the Partridge-colored, the Gray, or Brahmapootra of a few writers, the Dominique, and the Black.

About ten years ago there raged among our fowl fanciers a most alarming Shanghai fever. It had its "run," and its victims mostly survived. We presume they will never have a second attack.

We can not advise our readers to breed Shanghai fowls, and regret being obliged to mention them at all.

8. *The Cochin China Fowl, etc.*—A missionary in China says: "There is no difference at all between the Shanghais and Cochin Chinas. In reality they *all* are Shanghais. Cochin Chinese fowls are a *small, inferior* kind, not equal to the natives of the United States, and it is not believed that any have ever been taken to America;" and the editors of the "Poultry Book," lately published in London, quote from a letter they received from Mr. Robert Fortune, who has passed many years in various parts of China, as follows: "I firmly believe that what are called 'Cochin Chinas' and 'Shanghais' are one and the same.

Whether this testimony should be considered conclusive or not we leave the reader to judge, and believing none of the uncouth, awkward, and coarse-grained Asiatic fowls desirable, we herewith dismiss them.

9. *The Bantam Fowl.*—The Bantam is the smallest specimen of fowl, and may with propriety be called the Tom Thumb of the gallinaceous tribe, and stands comparatively, in size, to the Malay and Cochin fowl as that of the noble and stately Dur-

ham to the diminutive Alderney cow. Though extremely small in size, the Bantam cock is elegantly formed, and remark-

Fig. 37.

WHITE BANTAM COCK AND HEN.

able for his grotesque figure, his courageous and passionate temper, his amusing pompousness of manner, his overweening assumption and arrogance; and his propensity to make fight, and force every rival to "turn tail," has caused him many difficulties.

The Bantam must be considered more as an object of curios-

Fig. 38.

BLACK BANTAM COCK AND HEN.

ity than utility, and of course must expect to be received with no peculiar favor, in this country, except as a "pet." They

arrive at maturity early, are faithful sitters, good mothers, and will lay more eggs, though small, than any other variety. They are very domestic, often making their nests in the kitchen, depositing their eggs in the cradle or cupboard of the dwelling when permitted.*

The most beautiful of the Bantams is the Seabright, of which there are two sub-varieties—the Gold-laced and the Silver-laced.

The ground color of the Gold-laced should be a clear, golden, yellow-white; while in the Silver-laced it should be a pure silvery-white. The accompanying cut will give the reader a good idea of the form and bearing of these remarkable and beautiful fowls, as well as of the markings of their plumage.

The Seabright Bantam is emphatically the English gentleman's Bantam. Even lords and duchesses strive for the mastery in breeding this beautiful bird. This bird was first bred

Fig. 39.

THE SEABRIGHT BANTAM.

and introduced to the notice of English fanciers by the late Sir John Seabright, from whom they received their name.†

---

* Bement.   † Country Gentleman.

10. *The Game Fowl.*—The Game fowl is hardy, easily kept, and extra good for the table. The hens are fair layers, excellent sitters, exemplary mothers, and in every way well behaved

Fig. 40.

GAME COCK AND HEN.

fowls. The cocks have the reputation of being quarrelsome and tyrannical; but those who have studied their character most closely are of the opinion that, on this ground, they have been unjustly condemned. They are brave and powerful, but not pugnacious or vindictive. Bement says: "For those who do not wish to give much attention to fowls, there is, according to our opinion, no breed equal to the Game."

11. *Mongrel Fowls.*—The collections usually known under the name of Barn-door fowls or Dunghill fowls are merely rabbles of mongrels, in which the results of accidental or injudicious crosses have become apparent in all sorts of ways. There is a tendency among them to revert back to some one of the original breeds, and good fowls for all common uses are often found among them.

12. *Choice of Breed.*—We have mentioned the leading characteristics of the different kinds of fowls, in order to enable the reader to decide which is best adapted to his purpose. Were our advice asked in reference to the choice of a breed, we would recommend the Spanish where eggs are to be made the prin

cipal object, and the Gray Dorking where chickens are wanted for the table or for market. In reference to merely ornamental poultry, let "fancy" rule.

13. *Accommodations.*—No one should attempt to keep fowls without providing for them the proper accommodations to insure their comfort and health. These need not be expensive. A very simple house with appropriate accessories in the form of a yard, nests, feeding troughs, water basins or fountains, roosts, etc., can all be very cheaply furnished; or they may be more extensive, elaborate, and costly, if the proprietor's wants require and his means permit. For plans and descriptions of these structures we must refer the reader to "The House," which forms another number of this series of manuals. We need only say here that they should be such as to secure warmth and efficient shelter from storms, without excluding light or air, both of which are essential to the well-being of fowls as well as human beings.

"Most farmers," Mr. Bement truly says, "pay little or no attention to their fowls, suffering them to roam and run about when and where they please; to lay and hatch where it suits them best, and to roost on trees, under sheds, on the wagon, cart, hay-rigging, etc.—soiling by their droppings plows, harrows, or whatever may chance to be within reach. This treatment is no less unprofitable than inhuman. No wonder such farmers get no eggs during the winter, and generally come to the conclusion that poultry keeping does not 'pay.'"

Whatever may be the form or size of your poultry-house, it should be so constructed as to secure as equable a temperature as possible. This end is best attained by having the walls and roof lined, leaving an open space of from four to six inches between the outer and inner walls, which may be filled in with chaff, saw-dust, or dry tan. This will make it warm in winter and cool in summer. In addition to the inclosed portion, the house should have a broad piazza or shed attached, to which the fowls may retire for shelter in stormy weather.

Hens always seek to avoid observation when laying, and it is

well to gratify this natural feeling in the construction of their nests. A screen of lattice-work in front of the boxes, or a few evergreen boughs properly placed, will secure the required seclusion without preventing the circulating of the air.

In reference to the poultry-yard Mr. Bement says:

"Where it is intended to keep a large number of fowls, let the yard be of ample dimensions, which of course must be regulated by the number intended to be kept. Those contracted seven-by-nine pens which meet our eyes throughout the country are not calculated to answer the purpose for which they were intended. Half an acre, at least, for every hundred fowls (and more than that number should never be kept in one flock), is little space enough for them to roam in; and in order to unite all the advantages desirable in a poultry-yard, it is indispensable that it neither be too cold during winter nor too hot during summer; and it must be rendered so attractive to the hens as to prevent their laying in any chance place away from it. To shield them from the chilling blasts of winter and the scorching rays of the sun in summer, we would recommend planting evergreens on the borders of the yard, and shade trees in the center. This, with a good covering of grass, would leave little to be desired on that part. And if the fowls can have access to a grass field occasionally, and the soil *dry*, then, so far as the ground and situation are concerned, nothing to be wished for remains.

"A picket fence, from six to seven feet high, will be sufficient to prevent the fowls from flying over."

14. *Feeding.*—The fowl is as omnivorous as a pig or a man, and perhaps a little more so; nevertheless grain is their staple. Of this they ought to have a variety, as they do not thrive so well when fed constantly with one kind. Corn, wheat, barley, oats, and buckwheat make good feed for them. It is better to have all kinds of grain, intended for feeding fowls, and especially corn, coarsely ground or cracked. It will be found that they require a smaller quantity in this state. It should be scalded, or at least mixed to the consistency of a stiff batter

with water, before feeding it to them. Vegetables, such as potatoes, carrots, parsneps, beets, etc., boiled and mashed, are acceptable and wholesome. Lettuce, cabbage, Scotch kale, etc., chopped up fine, are excellent for all kinds of poultry in the winter. A few chopped onions may occasionally be added; and also a little flesh-meat, either raw or cooked, cut into small pieces.

The editor of the *Country Gentleman* thinks that it is better to feed poultry in winter from three to four times daily, than twice, which is the ordinary custom. By frequent feeding, the birds eat but a little at a time, and never injure themselves; but when fed but once or twice daily, there is danger of their overeating, which frequently produces fatal results. Our rule is, to so regulate the quantity given at each time, that each fowl shall have all it wishes, and have nothing left. Our experience confirms what many have said, that regular and frequent feeding is better for the health of the fowl, at any season of the year, than it is to fill a vessel with grain and allow them access to it at all times. We also think that poultry will eat less with frequent feeding than by twice feeding daily.

Lime is necessary for the formation of egg-shells, and should always be accessible. The best form is that of calcined oyster shells, pounded in small fragments. A box of sand and gravel, and another of ashes, should be added.

Pure water is another essential that can not be too strenuously insisted upon, impure water being a grand source of the diseases of poultry.

Cleanliness must be strictly attended to in all your arrangements for fowls; and the inside of the poultry-house should be whitewashed twice, at least, during the year, as a preventive against vermin.

15. *Incubation, and Rearing Chickens.*—For sitting, choose good-sized hens. Those with short legs, broad body, and large wings are best adapted to the duty. It is also generally remarked that the worst layers are the best sitters. All the eggs for a brood, which should not exceed thirteen, should be

so nearly as possible of the same age. None of them should be more than ten days old; and the reason why they should be of about the same date is, that they may be hatched simultaneously. Select eggs of average size and ordinary shape. Give the hen a quiet place to sit, and take care that she be not disturbed. In twenty-one days (sometimes a day or two earlier in warm weather) a good sitter will bring out the chicks. The first day after hatching they do not want food and should be left in the nest. The next day they may be put into a good coop in a dry, sheltered situation, and fed with coarse cornmeal mixed up with water, hard-boiled eggs chopped fine, or fresh curd. Feed a little at a time and often, and beware of overfeeding. When a little older, cracked corn, millet, wheat, barley, etc., may be fed to them. Have plenty of pure water in a shallow dish (so that they may drink without getting into it and wetting their feathers) always before them. After five or six days they may be allowed to range at will outside of the coop, but should not be allowed to come out while the dew is on the ground. When two or three weeks old, or, indeed, with the hardier breeds much earlier, the hen may be permitted to lead them out. The most important caution now is to guard them well against sudden unfavorable changes of temperature, and especially against cold rain-storms.

16. *The Poultry Pentalogue.*—Somebody in England has written a little work which he calls the "Poultry Pentalogue," in which the whole art of fowl-breeding is summed up in five grand rules:

1. Pure breed;
2. A constant infusion of fresh blood, and the careful avoidance of in-and-in breeding;
3. A varied diet;
4. Equable temperature; and—
5. Strict cleanliness.

Good rules and easily remembered. We commend them to our readers, who may profitably apply them to other stock besides fowls.

II.—THE GUINEA FOWL.

"There is no doubt," Wingfield says, "from the description given by Columella and Varro, that the Guinea fowl was reared on the farms of the Romans, and that it was first made known to them during their wars in Africa." They have hardly found the favor among poultry keepers that their merits would warrant. They are prolific layers of excellent eggs, and as table birds are by no means to be despised. They are shy, and love to make their nests in dark, obscure places, far

Fig. 41.

THE GUINEA FOWL.

from home; for which reason their eggs are generally placed under a common hen to be hatched and fostered. They give no notice of laying or sitting.

A brood of Guinea fowls is an excellent guard. They love roosting in the trees; and at night, if any footstep disturb them, their loud cries are sure to give notice to the farmer that a trespass is committing.

The Guinea fowl is delicate eating, and is in fine season about

Lent. The young chickens must be treated in the same manner and with the same food as young turkeys, and they must be kept warm and dry. In fatting, they should be shut up in a house for a fortnight, and fed four or five times a day with sweet barley-meal, moistened with milk and good lard. They pine if confined any length of time.

The great drawbacks to the rearing of Guinea fowls are the vigilance required to watch for their nest, and the harsh screaming of their cry.

III.—THE DOMESTIC TURKEY.

The domestic turkey is not so far removed from the wild state as the domestic fowl. There is no dispute about his origin, the wild turkey not being yet extinct, and not differing so widely from the tenants of our barn-yards as to give room for doubt on that point. In fact, as it is stated in the "American Poulterer's Companion," if kept in the neighborhood of large forests they will often stroll thither, without any design to return, such is the natural wildness of their species.

We have three varieties of the domestic turkey in this country—the Black, the Buff-colored, and the White. The Black is generally preferred, it being the most hardy. The Buff-colored is placed next in the order of merit. The White variety is very beautiful, but is smaller and less hardy.

Turkeys are generally considered very difficult to rear; and it is undoubtedly true that considerable care, patience, and skill are required to insure uniform success. Mr. Bement says: "If attempts to rear turkeys have not been crowned with success, it is entirely owing to the unskillfulness and inexperience of those to whom they have been intrusted; and so long as one persists in thwarting the females when sitting; in opening the shells of the eggs in order to help the passage of the tardy chicks; in pressing them, so soon as they are born, to eat against their will; and in leaving them exposed to intense heat, or to cold and dampness, so long will their death, in the course of a month, be the undoubted consequence. It is less trouble

to say the breed is difficult to rear, than to acknowledge at once that negligence, unskillfulness, and barbarity are the causes."

The principal requisites for the successful rearing of turkeys, according to the experienced author of the "Poulterer's Companion," are:

1. Good stock to breed from, both male and female. Both should be large and fully grown. They ought to be at least two years old.

2. Fresh blood, secured by changing the cock every year.

3. Good keeping through the winter.

4. No unnecessary interference with the process of incubation, which lasts four weeks.

5. Shelter, protection, and careful feeding of the chicks for a few weeks, after which the mother may be liberated from the coop to lead them out.

Curd chopped fine, crumbs of bread softened in water or milk, are good for their first food; but they will soon eat anything that is fit for the parent turkey, except unbroken grain.

Early in the fall they should be fed night and morning with dry corn; and when the weather becomes colder they may profitably be supplied at frequent intervals with boiled potatoes, mashed with corn meal and skimmed milk, given to them warm. On this diet they will grow and fatten rapidly.

The turkey is an out-door bird and requires, at most, only an open shed for shelter during severe storms, and even this will seldom be occupied if a good tree be at hand. They have not yet acquired all the effeminate artificial habits of the domestic fowl.

The critical periods with the turkey are about the third day after they are hatched, and when they have thrown out the "red head," as it is called, which they do when about six weeks old. To carry them safely through the first, avoid overfeeding, and secure them against unfavorable changes of temperature. In the latter case, give them a plenty of food, and render it as nutritious as possible by adding boiled eggs, wheaten grits, bruised hemp seed, or bruised beans.

Cobbett says: "As to fattening turkeys, the best way is to never let them get poor. Barley meal mixed with skimmed milk and given them fresh will make them fat in a short time. Boiled potatoes mixed with corn meal will furnish a change of sweet food which they relish much, and of which they may eat as much as they can."

IV.—THE DOMESTIC GOOSE.

The domestic goose has acknowledged the sway of man for ages—perhaps since the days of Noah. Homer mentions them, where Penelope, relating her dream, says: "I have twenty geese at home, that eat wheat out of water, and I am delighted to look at them." Their cackling, it will be remembered, saved

Fig 42.

EMBDEN OR BREMEN GEESE.

Rome from the Gauls, B.C. 388. Their wild original is unknown, the wild geese of the present day being of a different species.

Of the common domestic goose there is really but one variety divided into several sub-varieties, marked by more or less permanent distinctive characteristics—of these the Toulouse goose and the Bremen goose are probably the best. The former is gray and the latter white. The White China goose probably belongs to a distinct species. It is a beautiful bird, but comes properly under the head of ornamental poultry, of which we have little to say. It can be kept with advantage only in a warm climate.

Where there are facilities for keeping them, geese are considered the most profitable of all our domestic birds. The chief requisites for goose keeping are a pond or pool of water and a pasture for grazing.

The domestic gander is polygamous, but should not, Mormon-like, be allowed an unlimited number of wives. Three is sufficient, and some recommend to allow only two geese to each gander. Comfortable and well-ventilated apartments should be provided for geese, so constructed as to secure them against rats, weasels, skunks, etc. A separate room for the sitting goose is desirable. Her period of incubation is about thirty days. Thirteen eggs are the usual number given to the goose. She always covers them when absent from the nest.

"On the first day after the goslings are hatched," Mr. Bement says, "they may be let out, if the weather be warm. care being taken not to let them be exposed to the unshaded heat of the sun, which might kill them. The food given them is prepared with some barley or Indian meal coarsely ground, bran, and raspings of bread, which are still better if soaked and boiled in milk, or lettuce leaves and crusts of bread boiled in milk. On the second day a fresh-cut turf is placed before them, and its fine blades of grass or clover are the first objects which seem to tempt their appetites. A little boiled hominy and rice, with bread crumbs, form their food for the first few days, fresh water in a shallow vessel, which they can dabble in and out

without difficulty, being duly provided. Afterward advantage must be taken of a fine warm sun to turn them out on grass for a few hours; but if cold and damp, they should remain in their house, in which every attention should be paid to cleanliness by a constant supply of clean straw. After two weeks we cease these special precautions against exposure to the weather, and find them perfectly able to shift for themselves, in company with their mothers and the others of their race. For some weeks, however, extra supplies of food, such as bran or corn meal mixed with boiled or steamed vegetables, may be given them twice a day, morning and evening, continuing to give them this food till the wings begin to cross on the back, and after this, green food, which may be mixed with it, such as lettuce, cabbage, beet leaves, and such like. The pond is strictly forbidden them under all circumstances for the first two weeks, and in severer weather for a longer period. Exposure to heavy rain out of doors, and a damp floor in the house where they are placed at night, are the main hazards to be avoided."

One of the greatest sources of profit in goose keeping is the sale of the feathers; but plucking them from the living geese is a practice so full of cruelty that we can not conscientiously give any directions for the process. A writer in one of the magazines recommends *shearing* instead of plucking. He says: "Feathers are but of a year's growth, and in the moulting season they spontaneously fall off, and are supplied by a fresh fleece. When, therefore, the geese are in full feather, let the plumage be removed, very close to the skin, by sharp scissors, clipping them off as sheep are shorn; they will be renewed at moulting in the usual course of nature. The produce would not be much reduced in quantity, while the quality would be greatly improved, and an indemnification be experienced in the consciousness of not having tortured the poor bird, and in the uninjured health of the fowl, and the benefit obtained in the succeeding crop. After this operation shall have been performed, the down from the breast may be removed by the same means."

V.—THE DOMESTIC DUCK.

The origin of the tame duck is not a well settled point. Dixon supposes it to have been imported from India and China in or about the year 1493.

Of the numerous varieties known to the poulterer, Mr. Giles, of Woodstock, Conn., whom we have already had occasion to quote, recommends, for those who desire to keep ducks for use and not for ornament merely, the Rouen, the Java, and the Aylesbury.

The Rouen duck, originally from Rouen, France, is of a dark-brown plumage; legs and feet a dark dusky red; bill at the base black, tapering down toward the point a dark green, sometimes streaked with yellow; long in the body, with a small neck. The drakes are invariably the color of the wild Mallard drake, having a white ring around the neck; legs and feet a bright red; bill a bright yellow; flesh darker and higher flavored than the common duck. Very prolific, hardy, and easy to raise; will weigh at full maturity from eighteen to twenty pounds a pair.

The Java duck, originally from Java. Plumage a glossy black; neck long; round body; legs and feet black, and black bill. Drakes are black, head and neck bordering on a dark green; yellow bill; with bright red legs and feet. The Java ducks will attain to nearly the same weight as the Rouens—flesh similar.

The Aylesbury duck, originally from the town of Aylesbury, England. Plumage a beautiful white, with white bill; legs and feet a bright pink, ornamental in appearance; easy to propagate; producing white downy feathers, white skin, and delicate, savory flesh; will weigh from fifteen to eighteen pounds the pair. Sit the eggs under hens, and have them hatch out early. With care you can have large ducks.

The Wild Mallard duck is often domesticated. It is a very beautiful bird and becomes quite tame, rearing broods like the common duck; but no permanent tame race has yet been derived from them.

The Musk or Brazilian duck is from the tropical regions of South America. It is a singular bird in appearance and in habits, but we see little to recommend it, either for use or ornament.

The Wood duck, the most beautiful of its genus, so common in all parts of the North American continent, is also easily domesticated. It also will breed in its domesticated state.

Ducks are easily kept where there is access to a pond, pool, stream, or swamp. They will eat almost anything, animal or vegetable. The refuse of the kitchen garden is always acceptable to them, and where grass is not attainable, something of this kind must be regularly supplied.

"The duck-house," Bement says, "should, if possible, be of brick, and paved with the same material, with considerable inclination, so that the wet, when the floor is sluiced down, may at once pass off. Wood is seldom secure against rats, and does not so well suit the cleaning process of water and the lime-brush, and few places require their application more frequently. Do not crowd your birds, and always arrange for good ventilation. When the flock is large, separate the young ones, that they may thus have the advantage of better food, and that no risk may be incurred of finding the eggs of the older ones trodden under foot and broken at your morning visit. On this account the laying ducks should always have plenty of room, and be kept by themselves. Ducks, for these reasons, as well as for the sake of cleanliness, should never share the habitation of fowls, and from geese they are liable to persecution. Yet, where fowls are kept, a little contrivance will suffice to make their berth, even in a fowl-house, tolerably comfortable. In winter, a thin bedding of straw or rushes should be placed on the floor, and frequently changed."

The duck is a prolific layer, and her eggs are very rich and highly flavored, and are much relished by some persons. One duck's egg is considered of equal culinary value to two fowl's eggs.

According to Mr. Parmentier, one drake is sufficient for

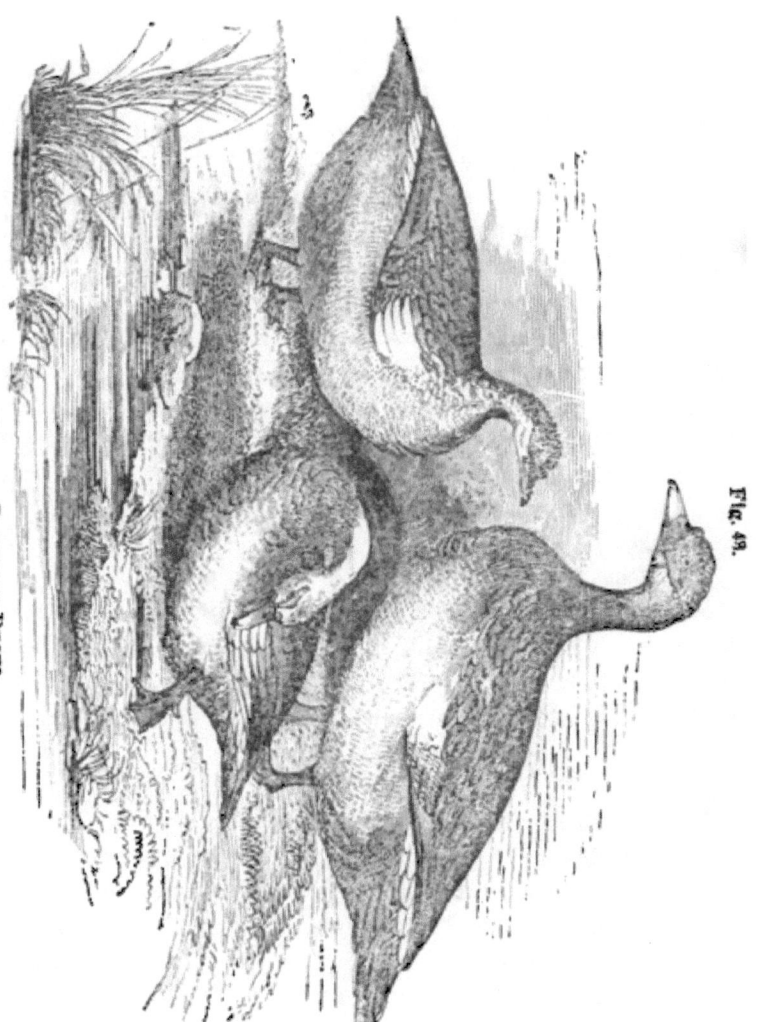

Fig. 49.

Musk or Brazilian Ducks.

eight or ten ducks, but others limit the number to from four to six.

Ducks are not so easily persuaded to lay in nests prepared for them, but prefer to choose a place out-of-doors to deposit their eggs. If the nest selected be tolerably secure, it is better to allow them to sit there than to attempt their removal. Thirteen eggs are a full allowance for a duck, and these should be as fresh as possible. The period of incubation varies considerably, but twenty-eight days is perhaps about the average time. The treatment of the young brood should be similar to that recommended for goslings. Boiled potatoes and hominy, or coarse corn meal, make excellent food for them. It is better to give them no uncooked food for several weeks after they are hatched.

To fatten ducks you must give them a plenty of good grain (corn and oats are to be preferred), and not allow them access to too much garbage. All fish and flesh, and especially putrid animal matter, of which they are fond, must be excluded from their diet, or the flavor of their flesh will be destroyed.

VI.—PREPARATION OF POULTRY FOR MARKET.

Messrs. Drew & French, extensive dealers in farm and market-garden produce, fruits, eggs, poultry, etc., 85 Barclay Street, New York, in answer to various inquiries addressed to them, carefully prepared and published, some time ago, the following directions, which should be as carefully followed by all who send poultry to the city markets and wish to get the highest price for it:

"*First*—Give no food for twenty-four hours previous to killing. Food in the crop is liable to sour, and always injures the sale. Purchasers object to paying for undigested food.

"*Second*—'Sticking' in the neck with a penknife is the best mode of killing. If the head is cut off, the skin recedes, and the neck bone looks repulsive.

"*Third*—Most of the poultry coming to this market is badly 'scalded' or 'wet picked.' 'Dry picked' is preferred, and sells

a little higher, other things being equal. Great care should be taken in picking to remove all the pin-feathers, and to avoid tearing the skin, particularly upon the legs, where it is most likely to be broken. If properly scalded, it looks best.

"*Fourth*—The intestines should not be drawn. After picking, the head may be taken off, and the skin drawn over the neck bone and tied. This is best, though much comes with heads on.

"*Fifth*—Next in order, it should be 'plumped,' by being dipped about two seconds into water nearly or quite boiling hot, and then at once into cold water about the same length of time. Some think the hot plunge sufficient without the cold. It should be entirely cold but not frozen, before being packed. If it reaches market sound without freezing, it will sell all the better.

"*Sixth*—For packing, if practicable, use clean hand-threshed rye straw. If this can not be had, wheat or oat straw will answer, if clean and free from dust. Place a layer of straw at the bottom of the box, then alternate layers of poultry and straw—taking care to stow snugly, backs upward, filling vacancies with straw, and filling the package so that the cover will draw down snugly upon the contents. Boxes holding not over 300 lbs. are the best packages.

"*Seventh*—Number the packages; mark the contents of each on the cover; place the invoice of the lot in one package, marked 'bill,' sending duplicate by mail; direct plainly to the consignee, placing the name of the consigner in one corner."

IX.

BEE-KEEPING.

Oh, Nature kind! Oh, laborer wise!
  That roam'st along the summer ray,
Glean'st ev'ry bliss thy life supplies,
  And meet'st prepar'd thy wintry day:
Go—envied, go—with crowded gates,
The hive thy rich return awaits;
Bear home thy store in triumph gay,
And shame each idler on thy way.—*Anon.*

I.—THE WONDERS OF THE BEE-HIVE.

THE accounts given, by naturalists and writers on bee-keeping, of the instincts and habits of the bee seem truly fabulous; and yet they are all founded on observation, and there seems to be no reason for calling them in question.

A hive of bees, we are told, consists of three kinds—females, males, and workers. The females are called queens, and only

Fig. 44.

THE QUEEN BEE.

Fig. 45.

THE DRONE.

one is permitted to live in the same hive; but one is essential to its establishment and maintenance. The males are called drones, and may exist in hundreds, or even thousands, in a

hive. The workers, or neuters, are the most numerous, and perform all the labor, collecting the honey, secreting the wax, and building the cells. The females and workers have stings at the end of the abdomen, but the drones have none. The queen lives in the interior of the hive, and seldom leaves it except to lead forth a swarm. If she be removed from the hive, the whole swarm will follow her. The queen is not only the governor, but also the mother of the community, she being the only breeder out of 20,000 or 30,000 bees, on which account she is loved, respected, and obeyed with all the external marks of devotion which human beings could give to a beloved monarch.

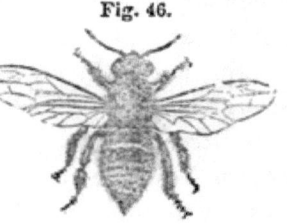

Fig. 46.

THE WORKER.

The queen deposits her eggs in cells previously prepared by the workers to receive them. The eggs producing workers are deposited in six-sided horizontal cells; the cells of the drones are somewhat irregular; those of the queens are larger than the others, circular, and hang perpendicularly. The eggs producing workers are laid first, the queen laying about two hundred eggs daily. The eggs of the drones afterward laid are less numerous than those of the workers, in the proportion of about one to thirty. Eggs for queens are deposited in their proper cells,

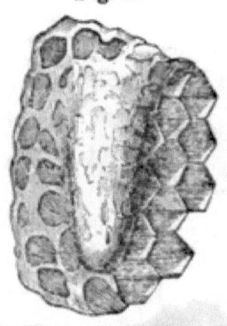

Fig. 47.

A ROYAL CELL.

one in each, at intervals of one or two days. The eggs and larvæ of the royal family do not differ in appearance from those of the workers, but the young are more carefully nursed, and fed with a more stimulating kind of food called "royal jelly," which causes them to grow so rapidly that in five days the larva is prepared to spin its web, and on the sixteenth day becomes a perfect queen. But as only one queen can reign in the hive, the young ones are kept close prisoners; and carefully guarded against the attacks of the queen mother

so long as there is any prospect of her leading out a swarm. When the old queen departs with a swarm, a young one is liberated, who immediately seeks the destruction of her sisters, but is prevented by the guards. If she lead forth another swarm, a second queen is liberated, and so on until further swarming is considered impossible, when the reigning queen is permitted to destroy her sisters. In cases where no new swarm is to be sent off, the queen mother is permitted to assume the office of destroyer. If at any time two queens happen to come out simultaneously, it is said that a mortal combat takes place at once, and the victor is acknowledged to be the rightful sovereign. On the loss of a queen, the whole swarm is thrown into the greatest confusion, and if there be no worker eggs or brood out of which a queen can be made by the peculiar process of feeding already mentioned, all labor ceases and the bees soon die.

There are three substances for which the bees forage the fields. First, a resin, or gum, which is on trees; next, the pollen, or fine dust, of flowers; and lastly, the saccharine matter that is in the flowers. When the cells are to be built, they bring home the resin, and stop all the cracks or crevices in the hive, so that neither the rain nor any insect can get in to trouble them. Then they set forth to bring materials for wax, to construct their cells. The wax is made from pollen. The bees swallow it, and then hang themselves in festoons in the hive. In the course of twenty-four hours small rings make their appearance on the body. Then the bee detaches itself from the rest of the group, and, descend-

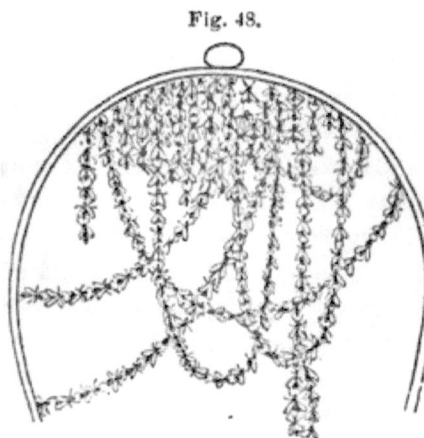

Fig. 48.

FESTOONS OF BEES SECRETING WAX.

ing to the bottom of the hive, removes the substance which has now become wax. Each bee follows in its turn, and deposits its contribution, which is directly made use of by the architects in building the cells.

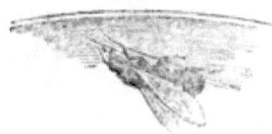

A WAX-WORKER.

The honey-cells are all six-sided, and of the most perfect regularity. Were they squares, or triangles, or circles, they would not fit as closely together, consequently there would be a waste of room.

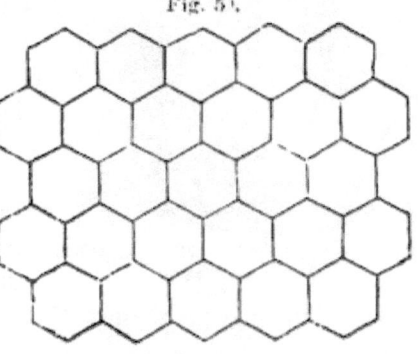

CELLS.

"There is a substance called bee bread, which is necessary to the life of the bee. It is made from pollen, but is entirely unlike wax. In securing it, the bee darts into a flower, and covers its body with the yellow dust. Now it must contrive some way to get rid of it, and God has made the last joint but one of each leg like a brush. These brushes are passed all over the body, and the pollen is collected in two little heaps. The thighs of the last pair of the bees' legs are furnished with two cavities, and these make nice little baskets to carry home their treasure. The dust collected from a thousand flowers is now kneaded into little balls, and when these have increased to the size of a grain of pepper, the bee flies home, and enters the cell head foremost. The balls are then taken from the baskets, and, being moistened with a little honey, become bee bread."*

Fig. 51.

BEE'S LEG MAGNIFIED.

---

* Student and Schoolmate.

One of the greatest wonders of the bee-hive is the mode in which it is ventilated. Fresh air is no less necessary to bees than to human beings, and as no provision is made for its supply in the construction of their dwelling, they secure it in this way: "They arrange themselves in files along the bottom of the hive. Those outside place their heads toward the entrance, and those within in an opposite direction. When thus stationed, they flap their wings so rapidly that we can not see that they have any wings at all. This rapid motion drives a current of air into the hive, to keep the honey and comb cool."

Fig. 52.

GLASS HIVE, SHOWING THE ARRANGEMENT OF COMB.

II.—THE APIARY AND HIVES.

The situation selected for an apiary or bee-house should be well sheltered from strong winds, and should not be near any large sheet of water. The hives should face the south, the east, or the southeast. They should be placed in a right line; and it is better to place them on shelves, one above another, than in rows upon the ground. The distance between the hives should be not less than two feet, and their height from the ground about the same. Near the apiary should be some small trees and shrubbery, on which swarms may alight; but large trees are objectionable. The grass should be frequently mowed around the bee-house, to prevent dampness and destroy the lurking-places of noxious vermin.

Much difference of opinion exists in reference to the best form and construction for a bee-hive, and many ingenious plans have been offered by the inventive genius of our country for their improvement. Some of these have peculiar excellences

and are worthy of a careful trial, but few if any of them are without some serious objections; so that practical bee-keepers generally prefer hives of the simplest construction. One of the best hives is made of pine boards an inch or an inch and a quarter thick. The best size is twelve inches square inside and fourteen inches deep. The top should be made of boards fifteen inches square. The boards should be joined carefully, and it is well to apply a coat of paint to the edges before putting them together. Small notches should be made at the bottom for the passage of the bees; and cross sticks put in for the support of the comb. If the inside of the hive be planed and covered with a thin coating of melted beeswax, it will save the bees much labor. Boxes for caps or covers may be fitted to these hives. These may be about seven inches deep and twelve square. They must fit closely the tops of the hives, and may be furnished with glass jars or other vessels for the reception of the honey. Several holes should be made in the top of the hive for the passage of the bees.

In Poland, where finer honey is produced and bees more successfully cultivated than anywhere else, the excavated trunks of trees are used for hives. Logs a foot or more in diameter and nine feet long are scooped or bored for the length of six feet from one end, the bore being from six to eight inches in diameter. A longitudinal slit is made in this hollow cylinder nearly the whole length and four inches wide. Into this slit is fitted a slip of wood with notches on the edges large enough to admit a single bee. This slip is hung on hinges and forms a door, by the opening of which the condition of the swarm can be seen and the honey be taken out. The top being covered, the trunk is set upright, with the opening toward the south. Sections of hollow trees are often used in this country for hives.

It is often desirable to carry honey to market without removing it from the hive in which it was made, and as few persons will purchase the contents of a large hive, one constructed in sections has a great advantage in that particular at least.

According to the views of Mr. Harasti, a skillful bee-cultivator, as quoted in the "Farmer's Encyclopedia," a good bee-hive ought to possess the following properties: First, it should be capable of enlargement or contraction according to the size of the swarm. Secondly, it should admit of being opened without disturbing the bees, either for the purpose of cleaning it from insects, increasing or dividing the swarm, etc. Thirdly, it should be so constructed that the produce may be removed without injury to the bees. Fourthly, it should be internally clean, smooth, and free from cracks or flaws. All these properties seem best united in the section-hive, which is constituted of two, three, four, or more square boxes of similar size as to width, placed over each other. Such hives are cheap, and so simple that almost any one can construct them.

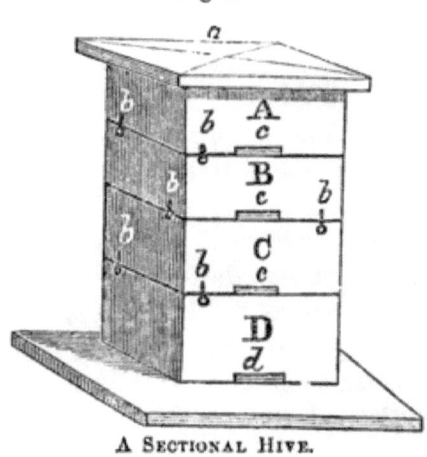

Fig. 53.

A SECTIONAL HIVE.

The boxes A, B, C, D may be made from ten to fourteen inches square and about five inches in depth, inside measure. Every bee-keeper should have his boxes made of the same size, so as to fit on to each other. Every hive must have a common top-board, *a*, which should project over the sides of the hive. The top-board of each section should have about sixteen holes bored through at equal distances from each other, and not larger than three fourths or smaller than four fifths of an inch. Or, instead of such holes, chinks of proper size may be cut through to allow the bees to pass up and down. At the lower part of each box or section, in front, there must be an aperture or little door, *c, c, c, d*, just high enough to let the bees pass, and about an inch and a half wide. The lowermost aperture, *d*, is to be left open at first, and when the hive is filled the upper

ones may be successively opened. By placing over the holes in the top of the upper section, glass globes, jars, tumblers, or boxes, the bees will rise into and fill them with honey. These may be removed at any time after being filled. The holes in the tops of the hive which do not open into the glasses or boxes should of course be plugged up. These glass jars, etc., must be covered over with a box, so as to keep them in the dark. Every box or section, on the side opposite the little door, should have a narrow piece of glass inserted, with a sliding shutter, by drawing out which the condition of the hive can always be inspected. To make the bees place their combs in parallel lines, five or six sticks or bars may be placed at the top of every section, running from front to rear. The bees will attach their combs to these bars, and the intermediate space will afford sufficient light to see them work. The slides covering the glasses should never be left open longer than is just necessary for purposes of inspection.

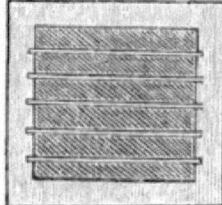

Fig. 54.

When one section is removed from the top, a wire or long thin knife must be previously run between this and the one immediately below, so as to destroy the attachments. Then remove the upper section, placing the top upon the one below, which is now the highest division of the hive. Another section is to be placed beneath, lifting up the whole hive for the purpose. Sometimes a second section has to be put under during a good season. If the swarm is not very large, three or even two boxes will be sufficient for its accommodation. The boxes or sections may be secured upon each other by buttons, b, b, or rabbets, and the joints closed with cement.

The plan of Mr. Luda, of Connecticut, by which the bees are made to build their cells and deposit their honey in the chamber of a dwelling-house appropriated for the purpose, in neat little drawers, from which it may be taken fresh by the owner, without killing the bees, has obtained deserved celebrity. The hive has the appearance of, and is, in part, a mahogany bureau

or sideboard, with drawers above and a closet below, with glass doors. This case or bureau is designed to be placed in the chamber of a house, or any other suitable building, and connected with the open air or outside of the house by a tube passing through the wall. The bees work and deposit their honey in drawers. When these or any of them are full, or it is desired to obtain honey, one or more of them may be taken out, the bees allowed to escape into the other parts of the hive, and the honey taken away. The glass doors allow the working of the bees to be observed; and it is said that the spaciousness, cleanliness, and even the more regular temperature of such habitations, render them the more industrious and successful.

III.—GENERAL DIRECTIONS.

1. *Swarming.*—Huish, in his "Treatise on Bees," says: "*The swarming of bees* generally commences in June; in some seasons earlier, and in cold climates or seasons later. The first swarming is so long preceded by the appearance of drones and hanging out of working bees, that if the time of their leaving the hive is not observed it must be owing to want of care. The signs of the second are, however, more equivocal, the most certain being that of the queen, a day or two before swarming, at intervals of a few minutes, giving out a sound a good deal resembling that of a cricket. It frequently happens that the swarm will leave the old hive and return again several times, which is always owing to the queen not having accompanied them, or from having dropped on the ground, being too young to fly to a distance. Gooseberry, currant, or other low bushes, should be planted at a short distance from the hives, for the bees to swarm upon, otherwise they are apt to fly away."

When they collect where they can not be shaken off and the hive can not be placed near them, they may be brushed off into a gauze sack, or any vessel in which they can be kept and carried to the hive, which should be set upon a table a little

raised on one side to allow their passage. If seen before they alight, they may often be secured by drawing a large woolen stocking upon the end of a pole and holding it up among them, as they are apt to consider it a favorable object on which to collect.

"When a hive yields more than two swarms, these should uniformly be joined to others that are weak, as, from the lateness of the season and deficiency in number, they will otherwise perish. This junction is easily formed, by inverting at night the hive in which they are, and placing over it the one you intend them to enter. They soon ascend, and apparently with no opposition from the former possessors. Should the weather for some days after swarming be unfavorable for the bees going out, they must be fed with care until it clears up, otherwise the young swarm will run great risk of dying."

Some recommend drawing off swarms without waiting for them to set forth of their own accord. We find the process thus described in the *Southern Homestead:*

"Those who are using a common hive when desiring to draw off a swarm, should let the hive be turned bottom upward, and the new hive set upon it; strike lightly upon the lower hive, and many of the bees will ascend into the upper hive; when a sufficient number has collected in the new hive for a swarm, take it off and set it upon the bench, and return the old one to its former position. In doing this, to insure success, it is necessary that one of the queens should accompany the new swarm, which may be known in the course of a day or two, for if they have no queen, they will not stay in the new hive, but will return to the old one; but if they have a queen, they soon manifest a disposition to commence work, and in the course of twenty-four hours some of the bees may be seen standing near the entrance of the hive, amusing themselves by raising their bodies to the full length of their legs, and giving their wings a rapid motion, making a steady buzzing noise. This may be considered as an indication of their satisfaction and the success of the operation. Some consider mid-

day the most favorable time for doing this; others again prefer the evening—but either will answer, and the trouble attending is not greater than that of hiving them when the swarms are allowed to come out in the common manner, and the danger of having them go off is avoided. Another very great advantage of this method is, the young swarms commence working early, by which they are more likely to lay up sufficient food for the winter."

2. *Robbing the Hives.*—The old practice, still followed by many, is to kill the bees by suffocation, whenever the most favorable time has arrived for taking the honey. To suffocate the bees, the hive is inverted over an empty hive or a hole in the ground in which some rags smeared with sulphur are burning. The bees drop down and are buried to prevent resuscitation. This is believed by some shrewd and experienced beekeepers to be the most profitable if not the most humane plan.

Polish apiarists cut out the comb annually to lessen the tendency to swarming, and thus obtain the largest amount of honey. In sectional hives it is readily taken out without killing the bees; and where these improved hives, as they are called, are not used, the comb may be cut out by merely stupefying the bees with sulphur or tobacco smoke. The time for taking up hives depends somewhat upon the season and pasturage; but the quantity of honey does not generally increase after the first of September.

3. *Wintering.*—To winter safely a swarm of bees, thirty pounds of honey are considered requisite. Only strong swarms are profitable to winter; therefore those that are found in the fall to be weak in numbers and with little honey had better be taken up. In the northern portions of the United States means are generally used to protect the swarms in winter, by removal to some cool and dry out-house or cellar; but many apiarists contend that this practice is not only useless but hurtful, and that hives should not be removed from their usual situations.

4. *Feeding.*—Bees are sometimes fed, when not able to supply their own wants, with a syrup made by dissolving brown

sugar in water and then boiling it to evaporate the water. Honey is the best food, but is generally (unless "Southern" or West India honey be used) too expensive; and, in fact, as a matter of profit, feeding should never be attempted.

5. *Killing the Drones.*—Knowing that the drones consume an immense amount of honey without producing any, and believing that a few of them will answer all the purposes required, Mr. P. J. Mahan, of Philadelphia, recommends getting rid of them, and thus saving the honey that they would consume. His plan for accomplishing this is to cut out the comb containing the cells in which they are to hatch. This, he says, is difficult in the common or box-hive and quite impossible in nearly all patent hives; but quite easy in Rev. L. L. Langstroth's Movable Comb Hive, in which the combs are built in a frame, similar to a slate or a picture in a frame, which being suspended on a narrow rabbet do not touch or come in contact with the hive at the top, bottom, or sides. Old combs can be put into the frames and be given to the bees to fill for their own use or for breeding combs.

"By cutting out the combs referred to," Mr. Mahan continues, "the bee-keeper makes a saving of all the honey fed to them before they are matured; the time occupied by the bees in feeding and nursing them; and last, though not least, assuming one foot as the average, which is capable of producing over 4,000 drones, by destroying this there is space sufficient to build combs in which 7,200 cells for hatching the workers will be erected; which, as we have done away with the drones, is fully equal to an accession of 14,400 working bees."*

This matter is certainly worthy of the attention of bee-keepers, and should be fully investigated.†

---

* Southern Planter.

† A large portion of the matter in this chapter, not credited to other sources, has been condensed from the excellent articles on "Bees and Bee-Keeping," in the "New American Encyclopedia."

APPENDIX.

HORSE-TAMING—RAREY'S SYSTEM.

1. THE THEORY.

THE one principle which you must establish firmly in your mind, and which is so essential in horse-taming that it is almost the corner-stone of the theory is the law of kindness. Next to kindness you must have patience, and next to patience indomitable perseverance. With these qualities in us, and not possessing fear or anger, we undertake to tame horses, with perfect assurance of success, if we use the proper means. The horse receives instruction in, and by the use of, four of his senses—namely, seeing, hearing, smelling, and feeling. You must remember that the horse is a dumb brute, has not the faculty of reasoning on experiments that you make on him, but is governed by instinct. In a natural state he is afraid of man, and never, until you teach him that you do not intend to hurt him, will that fear cease—we mean that wild, natural fear —for you must have him fear you as well as love you, before you can absorb his attention as much as is necessary to break him to your liking. It is a principle in the nature of a horse not to offer resistance to our wishes, if made known in a way that he understands, and in accordance with the laws of his nature.

In subjugating the horse, we must make a powerful appeal to his intelligence. This can only be done by a physical operation. It is an undisputed fact that the battles of all animals (except such as are garnished with horns) are fought by seizing each other by the throat. A dog that has been thus held by his antagonist for a few minutes, on being released, is often so thoroughly cowed that no human artifice can induce him to again resume the unequal contest. This is the principle upon which horse-taming is founded.

2. PRACTICAL RULES.

1. *Choking—First Method.*—Choking a horse is the first process in taming, and is but the beginning of his education. By its operation a horse becomes docile, and will thereafter receive any instruction which he can be made to understand. Teaching the animal to lie down at our bidding, tends to keep him permanently cured, as it is a perpetual reminder of his subdued condition.

It requires a good deal of practice to tame a horse successfully; also a nice

judgment to know when he is choked sufficiently, as there is a bare possibility that he might get more than would be good for him. We advise persons not perfectly familiar with a horse to resort rather to the strapping and throwing-down process (unless he is very vicious) described below; this, in ordinary cases, will prove successful. It is the fault of most people who have owned a horse to imagine that they are expert in his management; while, on the contrary, many professional horsemen are the very worst parties to attempt a subjugation. Unless a man have a good disposition, he need not attempt horse-taming.

In practicing the method exhibited in fig. 55, retire with the animal to be operated upon into a close stable, with plenty of litter upon the floor (tan-bark or sawdust is preferable). In the first place fasten up the left fore-leg with the arm strap, in such a manner that it will be permanently secured. Then take a broad strap and buckle and pass it around the neck just back of the jaw-bone. Draw the strap as tight as possible, so tight as to almost arrest the horse's breathing. The strap must not be buckled, but held in this position to prevent slipping back. The animal will struggle for a few minutes, when he will become perfectly quiet, overpowered by a sense of suffocation; the veins in his head will swell; his eyes lose their fire; his knees totter and become weak; a slight vertigo will ensue, and growing gradually exhausted, by backing him around the stable, he will come down on his knees, in which position it is an easy matter to push him on his side, when his throat should be released. Now pat and rub him gently for about twenty minutes, when, in most instances, he will be subdued. It is only in extreme cases necessary to repeat the operation of choking. The next lesson is to teach him to lie down which is described in the account of the fourth method of taming. No horse can effectually resist the terrible effects of being choked.

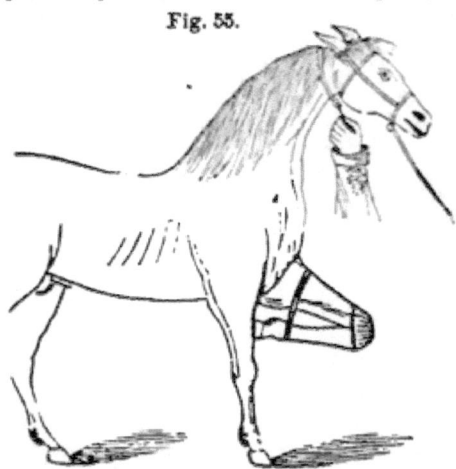

Fig. 55.

It must be constantly borne in mind, that the operator must not be boisterous or violent, and that the greatest possible degree of kindness is absolutely essential. When the horse is prostrate, he should be soothed until his eyes show that he has become perfectly tranquil.

2. *Second Method.*—The plan described in fig. 56 is very simple, though not as expeditious as the previous one. Buckle or draw a strap tight around the neck, lift a fore-leg, and fasten around it the opposite end of the strap, the shorter the better. In the engraving, for the sake of clearness, the strap is

represented too long. It will be seen that in this plan the horse is made the instrument by which the punishment is inflicted. When he attempts to put

Fig. 56.

his foot down, his head goes with it, and he thus chokes himself. Care should be taken that he does not pitch on his head, and thus endanger his neck.

3. *Third Method.*—Secure the horse with a stout halter to the manger. If extremely unruly, muzzle him. Sooth him with the hands for a few minutes, until he becomes somewhat pacified. Then seize him by the throat (as in fig. 57), close to the jawbone, with the right hand, and by the mane with the left. Now forcibly compress his windpipe until he becomes so exhausted that, by lightly kicking him on the fore legs, he will lie down, after which he should be treated as

Fig. 57.

previously described. This process requires courage in the operator, and also great muscular strength.

4. *Fourth Method.*—The horse to be operated upon should be led into a close stable. The operator should be previously provided with a stout leather halter; a looped strap to slip over the animal's knee; a strong surcingle, and a long and short strap—the first to fasten round the fore-foot which is at liberty,

and the second to permanently secure the leg which is looped up. The application of the straps will be better understood by reference to fig. 58.

In the first place, if the horse be a biter, muzzle him; then lift and bend his left fore-leg, and slip a loop over it. The leg which is looped up must be secured by applying the short strap, buckling it around the pastern joint and forearm; next put on the surcingle, and fasten the long strap around the right forefoot, and pass the end through a loop attached to the surcingle; after which fasten on a couple of thick leather knee-pads—these can be put on in the first place if convenient. The pads are necessary, as some horses in their struggles come violently on their knees, abrading them badly. Now take a short hold of the long strap with your right hand; stand on the left side of the horse, grasp the bit in your left hand; while in this position back him gently about the stable, until he becomes so exhausted as to exhibit a desire to lie down, which desire should be gratified with as little violence as possible; bear your

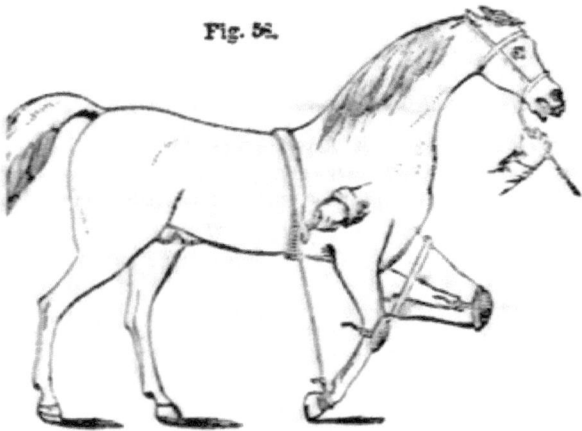

Fig. 58.

weight firmly against the shoulder of the horse, and pull steadily on the strap with your right hand; this will force him to raise his foot, which should be immediately pulled from under him. This is the critical moment; cling to the horse, and after a few struggles he will lie down. In bearing against the animal do not desist from pulling and pushing until you have him on his side. Prevent him from attempting to rise by pulling his head toward his shoulder. As soon as he is done struggling, caress his face and neck; also, handle every part of his body, and render yourself as familiar as possible. After he has lain quietly for twenty minutes let him rise, and immediately repeat the operation, removing the straps as soon as he is down; and if his head is pulled toward his shoulder it is impossible for him to get up. After throwing him from two to five times the animal will become as submissive and abject as a well-trained dog, and you need not be afraid to indulge in any liberties with him. A young horse is subdued much quicker than an old one, as his habits are not confirmed. An incorrigible horse should have two lessons a day; about the fourth

lesson he will be permanently conquered. If the operation is repeated several times, he can be made to lie down by simply lifting up his fore-leg and repeating the words, "Lie down, sir," which he must be previously made familiar with.

5. *Additional Hints.*—The following rules will serve as a guide to the amateur operator, and should be strictly observed:

First. The horse must not be forced down by violence, but must be tired out till he has a strong desire to lie down.

Second. He must be kept quiet on the ground until the expression of the eye shows that he is tranquillized, which invariably takes place by patiently waiting and gently patting the horse.

Third. Care must be taken not to throw the horse upon his neck when bent, as it may easily be broken.

Fourth. In backing him, no violence must be used, or he may be forced on his haunches, and his back broken.

Fifth. The halter and off-rein are held in the left hand, so as to keep the head away from the latter; while, if the horse attempts to plunge, the halter is drawn tight, when, the off-leg being raised the animal is brought on his knees, and rendered powerless for offensive purposes.—*New York Tribune.*

INDEX.

A.

	PAGE
Ass, The	45
Apiary	152

B.

Buyers, Hints to	42
Barns and Sheds	82
Breeds, Improvement of	108
Breeding In-and-In	109
" Hints on	112
Bee-Keeping	148
Bee-Hive, Wonders of	148
Bees, Swarming of	156
" General Management of	156

C.

Cattle, Breeds of	49
" Devon	51
" Hereford	53
" Sussex	54
" Ayrshire	54
" Welsh	55
" Irish	55
" Long Horn	56
" Short Horn	56
" Alderney	59
" Galloway	59
" Cream-Pot	60
" Points of	61
" General Management	65
" Weight of Live	71
Calves, Rearing	79
Crossing Breeds	110

D.

Diseases of Animals	114
Duck, The	143

E.

Ewe, Anecdote of	93

F.

	PAGE
Feeding Horses	31
" Sheep	84
" Cattle	66
" Swine	104
" Fowls	134
Fowl, The Domestic	118
" Spanish	119
" Dorking	121
" Polish	123
" Hamburg	124
" Dominique	126
" Leghorn	127
" Cochin China	129
" Bantam	129
" Game	132
" Mongrel	132
" Accommodations for	133
" Feeding	134
" The Guinea	137

G.

Guinea Fowl, The	137
Goose, The	140

H.

Horse, The	9
Horse, Breeds of	10
" The Racer	11
" Arabian	12
" Morgan	14
" Canadian	16
" Norman	16
" Cleveland Bay	18
" Conestoga	18
" Clydesdale	19
" Virginia	19
" Wild	20
" American Trotting	20
" Points of	21
" Color of	26
" How to Feed	31
" General Management of	35
" Vices and Habits	38
" How to Tame	161
Hives	152

L.

Lambs	88

M.

Mule, The	46
" Trade in Kentucky	47

P.

	PAGE
Piggery	107
Poultry	118
" Pentalogue	136
" Preparation of for Market	946

S.

	PAGE
Sheep, Breeds of	73
" Native	74
" Spanish Merino	74
" Saxon do.	76
" New Leicester	76
" South-Down	78
" Cheviot	80
" Lincoln	81
" Choice of Breed	81
" General Management of	82
" Value of to the Farmer	92
Swine, Natural History of	95
Swine. Opinions respecting	97
" Breeds	98
" The Land Pike	99
" Chinese	99
" Berkshire	100
" Suffolk	101
" Essex	102
" Chester	102
" Points of	104
" Feeding	104
Stables	29
Swarming	156.

T.

	PAGE
Turkey	138
Taming Horses	161

W

	PAGE
Water-Cure for Animals	118

WOODWARD'S
COUNTRY HOMES,

A New, Practical and Original Work on

RURAL ARCHITECTURE,

Elegantly Illustrated with

122 Designs and Plans of Houses of Moderate Cost,

INCLUDING STABLES AND OUT-BUILDINGS, WITH A CHAPTER ON THE CONSTRUCTION OF BALLOON FRAMES.

Price, $1 50 mailed free to any Address.

This work contains between its covers more practical information than can in many cases be sifted out of thousands of folios. The Messrs. Woodward are architects of note, and their work we warmly commend to our readers.—*Ithaca, N. Y., Journal.*

Contains a large number of very chaste and beautiful designs for snug, comfortable homes. The chapter on what is termed Balloon Framing, with the clear diagrams, is worth alone to country builders, the whole cost of the book.

We have long known these gentlemen as architects, and we regard them as among the most reliable and skilful men in the profession. Their new work on "Country Homes," ought to be in the hands of every man that builds or contemplates building a home.—*Scientific American.*

GEO. E. & F. W. WOODWARD,
PUBLISHERS,

37 Park Row, N. Y.

AGRICULTURAL PAPERS.

Files of the following papers can always be found in this office, and subscriptions received for them. It comprises, so far as we know, all the Agricultural and Horticultural periodicals in the country.

MONTHLY.

The Horticulturist—New York City,	$2.50
The Gardener's Monthly—Philadelphia,	2.00
Hovey's Magazine of Horticulture—Boston,	2.00
American Agriculturist—New York,	1.50
Maryland Farmer and Mechanic—Baltimore,	1.50
Working Farmer—New York,	1.00
Wisconsin Farmer—Madison, Wis.,	1.50
The Farmer (new) Richmond, Va.,	3.00
Sorgho Journal—Cincinnati, Ohio,	1.50
Southern Cultivator—Athens, Georgia,	2.00
Kansas Farmer—Lawrence,	1.50

WEEKLY.

Cultivator and Country Gentleman—Albany,	2.50
Prairie Farmer—Chicago,	2.00
Ohio Farmer—Cleveland,	2.50
Rural New Yorker—Rochester,	3.00
New England Farmer—Boston,	2.50
Boston Cultivator— "	3.00
Maine Farmer—Augusta,	2.50
California Farmer—San Francisco,	5.00
Iowa Homestead—Des Moines,	2.50
Western Rural—Detroit,	3.00
Germantown Telegraph—Germantown, Pa.,	2.50

SEMI-MONTHLY.

Coleman's Rural World—St. Louis,	2.00
Miner's Rural American—Clinton, N. Y.,	1.50

In addition to the above, we receive subscriptions to

Harper's Monthly Magazine	$4.00
" Weekly,	4.00
The Atlantic,	4.00
Our Young Folks,	2.00

And all other papers and periodicals published.

☞ Select your papers, remit us the amount by postal order, and the business will be transacted promptly.

GEO. E. & F. W. WOODWARD,
37 Park Row, New York.

ESTABLISHED IN 1846.

THE HORTICULTURIST

AND

Journal of Rural Art and Rural Taste,

DEVOTED TO THE

ORCHARD,
VINEYARD,
GARDEN,
AND
NURSERY,

To Culture under Glass, Landscape Gardening, Rural Architecture, and the Embellishment and Improvement of Country, Suburban and City Homes.

Published Monthly, and forming a handsomely Illustrated Annual Volume of 400 royal octavo pages, by the best practical talent in the country.

TERMS:

Two Dollars and Fifty Cents a Year—Twenty-five Cents a Number.

GEO. E. & F. W. WOODWARD,
PUBLISHERS,
37 Park Row, New York.

New and Revised Editions just published by Geo. E. & F. W. Woodward, 37 Park Row, N. Y.

THE HOUSE.

A NEW MANUAL of Rural Architecture; or, How to Build Dwellings, Barns, Stables and Out Buildings of all kinds; with a chapter on Churches and School-Houses. Cloth, $1.50.

THE GARDEN.

A NEW MANUAL of Practical Horticulture; or, How to Cultivate Vegetables, Fruits and Flowers; with a chapter on Ornamental Trees and Shrubs. Cloth, $1.00.

THE FARM.

A NEW MANUAL of Practical Agriculture; or, How to Cultivate all the Field Crops; with an Essay on Farm Management, etc. Cloth, $1.00.

THE BARN-YARD.

A NEW MANUAL of Cattle, Horse and Sheep Husbandry; or, How to Breed and Rear the various species of Domestic Animals. Cloth, $1.00.

Either of the above sent, post-paid, on receipt of price.

Agricultural, Horticultural and Architectural
BOOKS,

For Sale at Publishers' Prices at the Office of the Horticulturist, or mailed, post paid.

Orders executed for Books, Papers and Periodicals on any subject.

Grape Culture.

Chorlton on Grape Culture under Glass..	$ 75
Fuller's Grape Culturist...	1 50
Haraszthy Grape Culture, Wine and Wine Making......................................	5 00
Phin on Grape Culture...	1 50
Reemelin's Vine Dresser's Manual..	75

Fruit Culture.

Barry's Fruit Garden...	$ 1 75
Bridgeman's Fruit Cultivators' Manual...	75
Cole's American Fruit Book...	75
Downing's Fruits and Fruit Trees of America..	3 00
Eastwood on Cranberry...	75
Elliot's Western Fruit Grower's Guide..	1 50
Field's Pear Culture...	1 25
Fuller on Strawberry..	20
Hovey's Fruits of America, colored plates, 2 vols.....................................	35 00
Pardee on Strawberry...	75

Flowers.

Breck's Book of Flowers...	$ 1 50
Bridgeman's Florists' Guide..	75
Buist's Flower Garden Directory...	1 50
Ladies' Flower Garden Companion, edited by Downing..............................	2 00
Parlor Gardener...	1 00
Rand's Flowers for Parlor and Garden...	3 00
Skeleton Leaves and Phantom Boquets..	2 00
Wax Flowers, and how to make them..	2 00

Trees, &c.

Browne's Trees of America..	$ 6 00
Warder's Hedges and Evergreens..	1 50

Rural Architecture.

Allen's Rural Architecture...	$ 1 50
Cleveland's Villas and Cottages..	3 00
Cummings' Designs for Street Fronts, Suburban Houses and Cottages, with full exterior and interior details, 382 designs and 714 illustrations.....	10 00
Downing's Cottage Architecture..	2 50
Downing's Country Houses...	6 00
Hatfield's American House Carpenter...	3 50
Holly's Country Seats..	4 50
Leuchar's How to Build and Ventilate Hot-houses....................................	1 50
Manual of the House, 126 designs and Plans..................................cloth	1 50
Silloway's Modern Carpentry...	2 00
Sloan's Homestead Architecture, 200 Engravings...................................	4 00
Sloan's Ornamental Houses, 26 Colored Engravings................................	8 00
Vaux's Villas and Cottages, nearly 400 Engravings..................................	3 00
Woodward's Country Homes..	1 50
Woodward's Graperies and Horticultural Buildings..................................	1 50

Agricultural, Horticultural and Architectural Books.

Landscape Gardening.

Downing's Landscape Gardening	$6 50
Kern's Landscape Gardening	2 00
Kemp's " "	2 00
Rural Essays by Downing	3 00
Smith's Landscape Gardening	1 60

Gardening, Horticulture, Agriculture, &c.

Allen's American Farm Book	$1 50
Allen's Domestic Animals	1 00
American Rose Culturist	30
American Bird Fancier	30
Art of Saw-Filing	75
Bement's Rabbit Fancier	30
Bement's American Poulterer's Companion	2 00
Boussingault's Rural Economy	1 60
Boston Machinist, (W. Fitzgerald)	75
Brandt's Age of Horses, (English or German)	50
Bridgeman's Kitchen Gardeners' Instructor	75
Bridgeman's Young Gardeners' Assistant	2 00
Brown's Field Book of Manures	1 50
Buist's Family Kitchen Gardener	1 00
Burr's Field and Garden Vegetables of America	5 00
Canary Birds, Manual for Birdkeepers	50
Carpenters' and Joiners' Handbook	75
Cobbett's American Gardener	75
Cole's Veterinarian	75
Coleman's Agriculture	4 00
Darlington's American Weeds and Useful Plants	1 75
Dana's Muck Manual	1 50
Dana's Essays on Manures	30
Dadd's Anatomy and Physiology of the Horse..........Plain	3 50
Dadd's Horse Doctor	1 50
Dadd's Cattle Doctor	1 50
Davies Preparation and Mounting of Microscopic Objects	1 30
Farmers' Every Day Book, octavo, 650 pages	3 00
Flint on Grasses and Forage Plants	2 50
Flint on Milch Cows	2 50
Flora's Interpreter and Fortuna Flora, (Mrs. Hale)	1 50
French's Farm Drainage	1 50
Garlick's Treatise on Propagation of Fish	1 25
Gray's Manual of Botany	4 50
Guenon's Treatise on Milch Cows	75
Harris'—Insects injurious to Vegetation..........Plain Plates	4 00
" " " " ....Colored "	5 00
Harris' Rural Annual for 1866	25
Herbert's Hints to Horsekeepers	1 75
Hooper's Dog and Gun	30
How to Get a Farm, and Where to Find it	1 75
How to Write, Talk, Behave and do Business	2 25
Ik Marvel's Farm of Edgewood	2 00
Insect Enemies of Fruit Trees, (Trimble)	5 00
Jennings on Cattle	2 00
Jennings on Swine and Poultry	2 00
Jennings on the Horse and his Diseases	2 00
Johnston's Elements of Agricultural Chemistry	1 25
Johnston's Agricultural Chemistry	1 75
Klippart's Farm Drainage	1 50
Klippart's Wheat Plant	1 50
Langstroth on the Honey Bee	2 00
Liebig's Natural Laws of Husbandry	1 30
Liebig's Familiar Letters on Chemistry	50
Linsley's Morgan Horses	1 50

5

Agricultural, Horticultural and Architectural Books.

Manual of Agriculture, Emerson & Flint	$1 50
" of Flax Culture	50
" of Hop Culture	40
" of the Farm....................cloth	1 00
" of the Garden "	1 00
" of Domestic Animals "	1 00
Mayhew's Illustrated Horse Doctor	3 50
Mayhew's " Horse Management	3 50
Mayhew's Practical Book-Keeping for Farmers	90
Blanks for do do	1 20
McMahon's American Gardener	3 00
Miles on Horses Foot	30
Miss Hall, Cookery and Domestic Economy	1 50
Miss Beecher's Domestic Receipt Book	1 50
Miss Beecher's Domestic Economy	1 50
Morrell's American Shepherd	1 50
Munn's Practical Land Drainer	75
New Clock and Watch Maker's Manual	2 00
Norton's Scientific Agriculture	75
Onion Culture	25
Orchard House Culture, by C. M. Hovey	1 25
Our Farm of Four Acres, paper, 30 cents ; bound	60
Our Farm of Two Acres	20
Quinby's Mystery of Bee-keeping	1 75
Portfolio Paper File, (*Country Gentleman*)..........$1 and	1 50
Pedder's Land Measurer, for Farmers	60
Phenomena of Plant Life, (Geo. H. Grindon)	1 00
Randall's Fine Wool Sheep Husbandry	1 00
Randall's Sheep Husbandry	1 50
Ready Reckoner	50
Richardson, On Dogs	30
Rivers' Orchard House	50
Schenck's Gardeners' Text-Book	75
Shepherds' Own Book	2 25
Skilful Housewife	75
Stewart's Stable-Book	1 50
Saunders' Domestic Poultry...........paper 30c. cloth	60
Sparrowgrass Papers	2 00
Ten Acres Enough	1 50
Tenny's Natural History and Zoology	3 00
Thompson's Food of Animals	1 00
Tobacco Culture	25
Todd's Young Farmer's Manual	1 50
The Great West	1 00
Tucker's Annual Register of Rural Affairs, Nos. 1 to 12, each	30
Tucker's Rural Affairs, Four Bound Vols., each containing three numbers of the Annual Register, printed on larger and finer paper, per vol.	1 50
Turner's Cotton Planter's Manual	1 50
Waring's Elements of Agriculture	1 00
Watson's American Home Garden	2 00
Wet Days at Edgewood, by Ik Marvel	2 00
Wetherell on the Manufacture of Vinegar	1 50
Youatt on the Horse	1 50
Youatt on the Dog	2 00
Youatt and Martin, On Cattle	1 50
" On the Hog	1 00
Youatt, On Sheep	1 00
Youmans' Household Science	2 25
Youmans' New Chemistry	2 00

Address, **GEO. E. & F. W. WOODWARD,**
Publishers, 37 Park Row, New York.

WOODWARD'S
GRAPERIES
AND
HORTICULTURAL BUILDINGS,

By GEO. E. & F. W. WOODWARD, Architects & Horticulturists.

A new, practical and original Work on the Design and Construction of all classes of Horticultural Buildings, including

Hotbeds, Propagating Houses, Hot and Cold Graperies, Orchard Houses, Conservatories, &c.,

With the best modes of Heating, &c.

ELEGANTLY ILLUSTRATED.

Being the result of an extensive professional practice.

Price $1 50, Mailed Free to any Address.

"This neatly printed and finely illustrated work upon Horticultural Buildings gives full information upon the position and form of houses, manner of construction, heating, &c. Its plain directions for the erection and management of these structures, will command for it a wide sale, and being the result of the practical experience of well-known architects, its value as a hand-book to guide the novice will be highly respected.—*Maine Farmer*.

GEO. E. & F. W. WOODWARD,
PUBLISHERS,

37 Park Row, N. Y.

THE
DELAWARE GRAPE.

A MAGNIFICENTLY COLORED PLATE,

ON HEAVY ROYAL PAPER, FULL SIZE,

Being the finest thing of the kind ever Published in this Country.

Price per copy, mailed free, securely packed. **Three Dollars.**

GEO. E. & F. W. WOODWARD,
PUBLISHERS,

37 Park Row, New York.

www.ingramcontent.com/pod-product-compliance
Lightning Source LLC
Chambersburg PA
CBHW020300170426
43202CB00008B/446